ASTRONOMY

Reader's
Digest

The Reader's Digest Association, Inc.
Pleasantville, New York/Montreal

$\frac{26}{232} = 11\%$

CONTENTS

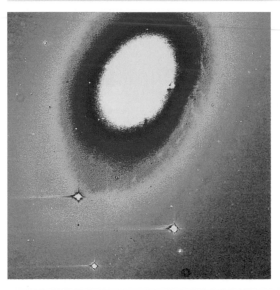

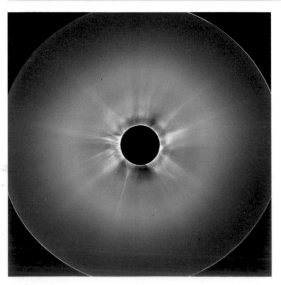

OBSERVING THE SKY 100

SPACE EXPLORATION 140

A Reader's Digest Book

Conceived and produced by Weldon Owen Pty Limited

A member of the Weldon Owen Group of Companies

The credits and acknowledgments that appear on page 160 are hereby made a part of this copyright page.

WELDON OWEN Pty Limited
PUBLISHER: Sheena Coupe
ASSOCIATE PUBLISHER: Lynn Humphries
PROJECT EDITOR: Jenni Bruce
ASSISTANT EDITOR: Gillian Hewitt
EDITORIAL ASSISTANTS: Vesna Radojcic, Shona Ritchie
ART DIRECTOR: Sue Rawkins
DESIGNER: Avril Makula
PICTURE RESEARCHER: Jenny Mills
ILLUSTRATORS: Nick Farmer, Chris Forsey, Robert Hynes
STAR CHARTS: Wil Tirion
INDEXER: Garry Cousins
PRODUCTION MANAGER: Caroline Webber
PRODUCTION ASSISTANT: Kylie Lawson
AUTHORS: Robert Burnham (Chapter 1, 3, 4, 5)
        Gabrielle Walker (Chapter 2)

Library of Congress Cataloging in Publication Data

Astronomy.
        p. cm. — (Reader's digest explores)
    Includes index.
    ISBN 0-7621-0042-7
    1. Astronomy—Popular works. 2. Outer space—
Exploration—Popular works.   I. Series.
QB44.2.A87 1998
520—dc21   97-46925

FROM SUPERSTITION TO SCIENCE

ONLY A FEW CENTURIES AGO, HUMANS
BELIEVED EARTH LAY AT THE CENTER OF THE
COSMOS AND THE MOVEMENTS OF HEAVENLY
BODIES INFLUENCED HUMAN DESTINIES. NOW
WE KNOW OUR PLANET IS BUT A MINUTE
PART OF A VAST, EXPANDING UNIVERSE THAT
FUNCTIONS ACCORDING TO PHYSICAL LAWS.

WHAT IS ASTRONOMY?

For as long as people have walked the earth, they have looked up with awe as they tried to understand the tiny points of light in the night sky. Today's astronomy has made enormous strides since humanity's first attempts. But in one respect the study of the heavens hasn't changed since the dawn of the Ice Age: despite its "otherworldly" focus, astronomy remains a central part of human culture.

No other science gets to grips with such big-picture questions. How did the universe form? Where did the stars come from? What

BELOW: Pillars of gas and dust reach a light-year high in the Eagle Nebula. In this and other views, the Hubble Space Telescope continues to revolutionize astronomy.

RIGHT: *The planets and other bodies of our solar system lie within reach of our spacecraft and, in some cases, of astronauts as well. The* Cassini *mission to Saturn, continuing missions to Mars, and bold new ventures to comets and asteroids are helping us better understand our planetary neighborhood and how Earth fits into it. Here,* Titan IV, *the launch vehicle for the* Cassini *craft, lifts off.* Cassini *is due to arrive at Saturn in 2004.*

BELOW: *One of the most powerful ideas in history was the discovery that the universe behaves like a giant clockwork, following forever a set of physical laws that can be understood by human reason. In an era before planetariums, clockmakers built models called orreries to show the movements of the Moon and planets throughout the year. The orrery shown here dates from about 1740.*

will happen to the solar system? And perhaps the biggest one of all: Are we alone in the universe?

The answers to these questions are not final. Like any science, astronomy is more a way of investigating the universe than a conclusive list of facts about it. Astronomy can never be complete because our quest to know how the universe works will never reach an end.

NEW WORLDS

One of the most exciting things about the astronomy described in this book is how new it all is. Quite literally, much of it could not have been written until the last few years. This largely reflects the current generation of telescopes—both on Earth and in orbit around it. These instruments have gathered views of the universe that are compelling astronomers once again to rewrite the book of the cosmos.

To pick just one new discovery—we now know of about a dozen planetary systems besides our own. Such systems have long been suspected, but their existence is now a verified fact—and many of them turn out to be quite unlike our own solar system. As more extrasolar planets are discovered, we'll gain a much better sense of how special our own system is.

And within the Sun's family, all planets (except Pluto) have received spacecraft visits. These explorations have shown us Martian volcanoes dwarfing any on Earth, for instance, and curious "racetrack" patterns on Miranda, a moon of Uranus. In just a few decades, the planets have changed from mysteriously moving lights in the sky into places that have geographies and histories as real as Earth's.

THE BACKYARD UNIVERSE

Closer to home, backyard skygazing has also changed dramatically. New books and charts, improved binoculars, and sophisticated telescopes have made astronomy a much more accessible hobby than it once was. Now anyone is able to step outside on a clear night and roam the universe as far as the eye and imagination can go.

If you want to experience the thrill of exploring the night sky, it has never been easier. Astronomy clubs have sprung

up throughout the world. Many clubs sponsor observing get-togethers, known as star parties, in dark-sky locations. At star parties, you can meet others interested in seeing the heavens—by eye, binocular, or telescope.

INTO THE NEW MILLENNIUM

When we look back four centuries to Galileo's time, we can feel wonder and excitement in his writings. He's seeing a new universe—the Copernican one—unfolding before his new-made telescope. Astronomy today stands on the threshold of the millennium. Now as then, new instruments and new ways of looking are bringing us new and better views of the extraordinary universe that surrounds us.

We're lucky enough to be living in another Galilean age—one that our descendants will look back on just as we look back on Galileo's. And the new astronomy that will emerge from our age will surely be as revolutionary as the discoveries that came out of his.

AGES OF MYSTERY AND WONDER

It's probably impossible to recapture how the night sky looked to people at the dawn of civilization. The barriers are both physical and mental. The fog of light that covers our cities blots out all but the Moon and the brightest stars and planets. Oblivious to the sky above us, we pursue our daily lives and leave studying the heavens to astronomers, who indeed have much to tell us from their remote observatories. Yet for all today's knowledge, the Mesopotamian shepherd of 6,000 years ago knew his heavens far better and more personally than we do ours.

The lives of ancient peoples were tied closely to the natural world, defined by the daily arc of the Sun and its year-long course through the seasons. In high latitudes (then as now), winter's deadly grip alternated with summer's plenty, while in the tropics seasons connoted cycles of dry and wet. The Moon provided light for hunting, either animals or enemies, and its 29½-day journey around the sky—one "moonth"—made a useful clock.

PATTERNS IN THE NIGHT SKY

No one knows where or when astronomy began. In Europe archeologists have found what may be lunar calendars that were carved into bone 30,000 years ago, and prehistoric solar-aligned monuments survive in both the New and Old Worlds. Neolithic people in northern latitudes watched the night sky intently, and were guided by its changes as they followed game animals or gathered the first crops. In fact, ice-age people probably inherited some sky learning from earlier paleolithic times.

As a species, human beings characteristically look for patterns in events and the things that surround them. Seeing people and powerful deities in the stars would come naturally, and the figures would act as memory aids for tribal lore at a time when all culture and learning lived in people's memories alone.

THE ZODIAC AND OTHER CONCEPTS

The first constellations likely marked places in the sky that seemed important. Foremost is the zodiac, the apparent yearly path of the Sun among the stars. By noting which bright stars rose just before the Sun, early observers saw that the Sun moved along a certain band of constellations. Now called the zodiac, this was also where eclipses occurred—rare, terrifying events in which the Moon turned copper, or the Sun was shut off for what must have felt like an eternity.

Five stars moved along the zodiac like the Moon did, but more slowly and less brightly. Each was home to a certain god. Perhaps because their movements were a mystery until only 400 years ago, these stars are still known by the Greek word for wanderer: *planetes*.

Beyond the zodiac, other mythological figures populated the skies. The earliest constellations have long since vanished, and the names we use are almost wholly Greek as filtered through Rome. Nor is there

ABOVE: The primary god in many traditional cultures is the Sun god, shown here in a wooden mask from the Pacific Northwest of America. A male Sun god was usually paired with a female Moon god, who ruled the night as he did the day.

anything special about the figures we see in the sky—for instance, the Chinese, Native Americans, and Australian Aborigines often grouped the same stars in different ways.

As bands of people ceased wandering and built agricultural settlements, anxious farmers eyed the stars for signs that the gods were pleased. With the rise of the first city-states about six or seven thousand years ago in Mesopotamia, the first astronomer-astrologers appeared—priests charged with tracking the heavens to learn the gods' intent.

The First Astronomers

Early records that are clearly astronomical first appear in Mesopotamia, the region that is now southeastern Iraq. They include a list of constellations of the zodiac compiled by the Assyrians about 700 BC, roughly the time of Homer. The list contains observations that probably date from several centuries earlier and, interestingly, uses Sumerian, not Assyrian, names for the stars and constellations.

The Sumerian Legacy

Sumeria was the first great civilization of the Middle East. In the fourth millennium BC, it sprang from agricultural city-states along the Tigris and Euphrates rivers. The Sumerians devised the plow, wheeled vehicles, major irrigation projects, and writing—the latter before about 3000 BC.

In time the Sumerians were supplanted by the Babylonians and Assyrians, but their skylore endured. Babylonian and Assyrian astronomy developed calendars for planting crops, and charted the Moon's phases and the apparitions of Venus. The Assyrians knew all five naked-eye planets, believing that each controlled events on Earth. In Assyrian astrology, Mars and Mercury brought misfortune, while Jupiter signaled good luck.

From Mesopotamia, much astronomical knowledge passed to the Greeks (see page 14). Familiar constellations such as Auriga, Scorpius, Sagittarius, Capricornus, Leo, Taurus, and

many others descend to us from our Sumerian forebears, with only a change of name to mark their passage through intervening cultures.

Predictions and Omens

Other ancient civilizations also developed astronomy. The lives of the Egyptians were shaped by the Nile River. To predict the river's floods, which fertilized and irrigated their crops, they watched for the rising of the star Sirius just before the Sun. They also drew their own constellations, depicting our Little Dipper as a hippopotamus, and the Big Dipper as a bull.

By the fourth century BC, Chinese astronomers had determined the length of the year as 365.25 days. But as the emperor formed a personal link between heaven and Earth, Chinese astronomy ran heavy with divination, seeking the right ritual to avert a bad omen or enhance a good one. Although astrology often took the upper hand, Chinese astronomers were accurate observers, compiling useful records of comets, supernovas (in particular that of AD 1054 in Taurus, which created the Crab Nebula), and solar and lunar eclipses.

LEFT: Astrolabes, such as this ancient Assyrian relic, were portable instruments that allowed the user to determine the time of night by measuring the elevations of selected stars in the sky. Astrolabes were used for thousands of years, and out of them evolved instruments such as the mariner's sextant.

FOLLOWING THE HEAVENS

In the islands of Oceania, astronomy revolved around navigation. Stars and planets rise and set at steep angles to the horizon in the tropics. Thus a bright celestial object makes a beacon that a sailor can steer by for several hours without going seriously off-course. Aided by stick-and-shell maps, Polynesian navigators of a thousand years ago memorized sky stories. These tales conveyed sailing directions using specific stars and constellations.

In the New World, hardly any written astronomy survived the conquests by Europeans. Only recently have archeologists learned to read the stone inscriptions left by the Maya of southern Mexico, who flourished between the third century BC and the ninth century AD. Mayan astronomers built a complex 52-year calendar around cyclic apparitions of Venus, which they associated with the god of rain.

The Aztecs dominated central Mexico for two centuries before the Spanish conquest in AD 1520. Little astronomical lore has survived, but we know that Venus was seen as the god Quetzalcóatl. This feathered serpent symbolized the power of life emerging from earth, water, and sky. It required rituals and sacrifices

as it disappeared and reappeared over a cycle of five apparitions in a span of eight years.

Farther north, in today's Wyoming, is a ring of stones called the Big Horn Medicine Wheel. While its exact function, builders, and construction date remain unknown, the spokes of the wheel are oriented toward the summer-solstice sunrise and sunset. They may also mark sightlines to the points on the horizon where stars rise.

EARLY MODELS OF THE UNIVERSE

LEFT: The idea that Earth lay at the center of the universe gave astronomy a comfortable framework for thousands of years, and it is still embedded in our culture—who doesn't speak of the Sun "rising" or "setting"? In this 16th-century Italian manuscript of the universe, around Earth lie nested spheres carrying the Moon, the Sun, the five pre-telescopic planets, and the stars, the last surrounded by the figures of the zodiac.

RIGHT: Pythagoras (near right) believed that the universe consisted of crystalline spheres, a notion developed further by Aristotle (center), whose methods of reasoning drew more on logic than experiment. His Earth-centered astronomical views reached their finest development in the work of Ptolemy (far right), whose system of mathematical astronomy held sway for some 1,500 years.

Greek astronomy often rings familiar—even the word *astronomy* is Greek and means "the naming of stars." The earliest Greek writers to mention astronomical matters are the poets Homer and Hesiod, who lived about 750 BC. The Pleiades, Orion, Taurus, Boötes, Ursa Major, and the star Sirius were all part of the night sky to Achilles and Odysseus. And Hesiod's rustic poem *The Works and Days* provided an agricultural calendar that was based on the visibility of constellations and stars. (For example, the best time to prune one's grapevines is when Arcturus rises from the ocean at sunset.)

THE GREEK PHILOSOPHERS

But the main thrust of Greek astronomy was philosophical—that is, it sought to describe the world. The first thinkers to speculate about the universe, free of pressure from rulers' ideologies or "practical" astrological concerns, were the Greek colonists living in Ionia (now the Asian coast of Turkey) and southern Italy. The Ionians also benefited from contact with Mesopotamian and Egyptian astronomy and mathematics. Regrettably, the earliest philosophers are now just names with only approximate dates. None of their works survives except in quotes or descriptions by later writers.

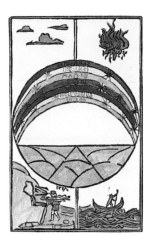

ABOVE: *In the four physical elements posited by the Greek philosopher Empedocles—air, fire, earth, water—we see ancient philosophy beginning its struggle to find rules that govern the everyday world. As air and fire weigh little, he thought they should dominate the sky, with the heavier earth and water forming the ground.*

The foremost Ionian was Thales of Miletus (sixth century BC), who is said to have predicted a solar eclipse. He also worked out the length of the year and knew the movement of the Sun through the zodiac.

Pythagoras (sixth century BC) was another Ionian, better known for his theorem on right-angled triangles. He proposed the universe was made of concentric spheres around a central fire, of which the Sun was merely a reflection. An opaque counter-Earth blocks our view of this fire, he said, while the Moon reflects the Sun's light. The Sun, Moon, and planets all had their own spheres separate from the one carrying the stars. (He also taught that as the spheres rubbed against one another they produced celestial music.) To account for known irregularities of motion, Eudoxus of Cnidus (fourth century BC) took the Pythagorean spheres and added to them, allowing a greater range of movements.

The great Greek scientist Aristotle (fourth century BC) correctly explained lunar phases and eclipses, and showed that Earth is spherical. However, he chose an Earth-centered universe because he could not see the stars' relative positions changing during the year, as they should if Earth were orbiting the Sun and the stars lay at different distances. (In fact, the positions do change, but so minutely that only sophisticated instruments can detect the differences.)

MEASURING THE UNIVERSE

In the third century BC, Aristarchus of Samos proposed a Sun-centered model, and tried to measure the distance to the Sun, but his ideas failed to dislodge the geocentric model.

A century later, Hipparchus built an observatory on the island of Rhodes and compiled the first known star catalog. He apparently adopted many methods and observations from Babylonian astronomers. Hipparchus devised our present magnitude system for measuring star brightnesses (see page 105), and measured the distance to the Moon as 29.5 times Earth's diameter (the correct figure is 30). Hipparchus also compared old observations with those of his day and discovered that the position in the sky of the north celestial pole had changed, because of what is now called precession of the equinoxes.

The last and greatest Greek astronomer lived in the second century AD in Alexandria: Claudius Ptolemaeus, called Ptolemy. In a work that is best known by its Arabic title of *Almagest*, Ptolemy presented a complete mathematical description of the movements of the Sun, Moon, stars, and planets—all using an Earth-centered model. His book was enormously influential, and shaped the course of astronomy until the Renaissance, when it was replaced by Copernicus's heliocentric model.

THE BIRTH OF MODERN ASTRONOMY

ABOVE: The Copernican solar system (seen here in a 17th-century Dutch engraving) was an act of faith at the time it was proposed. There was no observational evidence to support a heliocentric universe until long after Copernicus died.

Fourteen centuries separate Ptolemy's *Almagest* (about AD 150) from *On the Revolutions of the Celestial Spheres* (1543) by Copernicus. During this time, astronomy advanced little. Arab astronomers translated and studied Ptolemy's work (giving it the title it is known by), and skillfully refined it using accurate observations made with improved instruments. But they broke no new ground.

The reasons for European astronomy's long sleep are not difficult to find. Consolidating its power in the face of barbarians and Islam, the Christian Church essentially canonized Ptolemy's Earth-centered model of the universe. With that issue put to rest, the Church turned its main efforts to protecting and expanding control over theology, politics, and learning.

About AD 1000, European Christendom began to import some of antiquity's knowledge from the Muslim, and, later, the Byzantine, world. By the time of Dante (14th century), western astronomy had regained the importance it had held more than a thousand years earlier at the death of Ptolemy. One of those drawn

RIGHT: *Two telescopes that shook the world. The long instrument is a refractor built by Galileo Galilei and used to discover craters on the Moon and the moons of Jupiter. The other is Isaac Newton's reflector—the forerunner of many of today's telescopes, both professional and amateur.*

ABOVE: *The quadrant was an early instrument used to measure the elevations of stars and planets. This brass quadrant was made in 1549.*

RIGHT: *While Galileo did not invent the telescope, he was the first to observe the heavens with one and write about what he saw. The pages of his 1610 book* The Starry Messenger *record his intense excitement in discovering things that had never been seen before.*

to the newly invigorated astronomy was Nicolaus Copernicus (1473–1543) of Poland. Copernicus is frequently depicted as more radical than he really was. His great work, *On the Revolutions of the Celestial Spheres* (1543), retained many features of Ptolemy's universe. Its one significant difference: Copernicus placed the Sun at the center and demoted Earth to the status of a spinning planet.

FROM COPERNICUS TO KEPLER

Copernicus's ideas slowly spread. A number of philosophers were attracted to the new cosmology while others rejected it. Among the latter was a Danish nobleman named Tycho Brahe (1546–1601). Tycho Brahe was the finest astronomical observer the world had seen, and he compiled an enormous number of accurate observations with the goal of re-establishing astronomy from first principles. To that end, he devised a complex cosmology that included aspects of Ptolemy's model and aspects of Copernicus's. In Tycho Brahe's cosmology, the planets circle the Sun, but the Sun circles a stationary Earth.

Shortly before his death, Tycho Brahe took on an assistant, the Austrian mathematician and astronomer Johann Kepler (1571–1630). Inheriting Tycho's meticulous observations, Kepler (a thorough Copernican) used them to discover that planets orbit in ellipses, rather than in circles, and that their orbits and periods of revolution around the Sun follow a simple law (see pages 60–61).

THE WORK OF GALILEO AND NEWTON

Contemporaneous with Kepler, Galileo Galilei (1564–1642) of Italy used the newly invented telescope to investigate the heavens. He found craters and mountains on the Moon, the four largest moons of Jupiter, sunspots, the phases of Venus, and the starry nature of the Milky Way. His discoveries, plus his Copernican views on cosmology, led to a conflict with the Church, however. As a result, Galileo was threatened with torture by the Inquisition and put under house-arrest for the rest of his life.

The man who placed Copernican astronomy on a solid footing was born in England the year Galileo died. Isaac Newton (1642–1727) was one of the greatest scientists ever. His book *The Mathematical Principles of Natural Philosophy* laid the cornerstone for all physical science since his time. It presented the universe as an immense, impersonal machine, moving with a clocklike regularity that could be understood and predicted.

Newton's work transformed astronomy, which abandoned the purely mathematical hypotheses developed by the ancient world and henceforth dealt only with real, physical causes.

ASTRONOMY TRANSFORMED

After Newton, astronomers devoted almost all their efforts to analyzing the celestial clockwork in ever-greater detail. The discovery of the planet Uranus in 1781 was accidental, but the discovery of Neptune in 1846 followed a careful use of Newtonian celestial mechanics.

SPECTROSCOPY AND ASTROPHYSICS

For all their skill in describing celestial motions, early 19th-century astronomers still had little idea what stars and planets actually were. That began to change with the invention of the spectroscope in 1814 by Joseph Fraunhofer, who used it to find dark lines (corresponding to chemical elements) in the spectrum of the Sun. In years following, Robert Bunsen, Gustav Kirchoff, William Huggins, and many others built spectroscopes to study the planets and the brightest stars. Simultaneously, laboratory scientists began to analyze and classify multitudes of spectral lines produced by chemical elements, molecules, and compounds.

These studies showed the Sun and stars were balls of hot, luminous gases, while planets were inert reflectors of sunlight. When astronomers turned to view the "nebulae" in the sky, some had Sun-like spectra and were probably collections of stars; others showed only the bright emission lines of one or two elements and were clearly thin clouds of gas. In 1868 Norman Lockyer discovered a new substance in the spectrum of the Sun; he called it helium. Later, it was discovered on Earth.

At the end of the 19th century, the human eye was eclipsed by the photographic plate, and the spectrograph became the main tool of a new kind of astronomer: the astrophysicist. This change in name reflected the increasing overlap of astronomy with physics, out of which grew new tools such as quantum theory and, a little later, the theory of relativity.

By the 1930s, astrophysicists understood that stars came in many sizes and temperatures and differed in chemical composition. Researchers were beginning to build theoretical models describing how stars form, evolve, and die. After World War II, the first electronic computers enabled astrophysicists to build more detailed models. These were checked against observations that looked for ever more subtle details. Feedback between theory and observation soon brought astrophysics to maturity.

LEFT: *Albert Einstein (far left) developed the theory of relativity, which proved highly important to the emerging observational science of cosmology, largely founded by Edwin Hubble (left). When Hubble and other astronomers sought to understand the structure of space-time in the far reaches of the universe, they found that Newtonian physics fell short. Only relativistic equations had the power to explain what happens in regions of space where ordinary geometry breaks down.*

RIGHT: From the 1840s, the human eye was gradually replaced by photography, and new discoveries emerged. While early efforts centered on the Moon and Sun, astronomers soon turned their attention to more distant objects. This photo of the Pleiades star cluster was taken in the early 1900s.

BELOW: A forest of dark lines crosses the visible spectrum of the Sun, each line produced by atoms in the Sun's atmosphere that absorb light at specific wavelengths. By studying such lines, astrophysicists can determine what a star is made of, how hot and how bright it is, and many other properties.

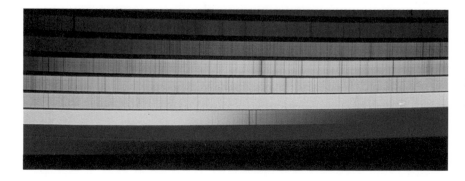

PLANETARY STUDIES AND COSMOLOGY

By contrast, there was little ferment in planetary studies. Despite the discovery of Pluto in 1930, this field remained a backwater that was visited by only a handful of scientists. After broadly determining the nature of planetary atmospheres and surfaces, there was limited progress in solar-system science until the space age sent instrumented probes to view these bodies at close range. Once these developments took place, however, the field blossomed quickly.

(As an indication, astronomy textbooks from the 1950s deal with the solar system in about 25 percent of the pages; today, the subject takes up half the book or more.)

Another changed field is cosmology, the study of the origin and structure of the universe, which began the 20th century as a forgotten corner of astronomy. Through his discovery of variable stars (see page 36) in the Great Galaxy in Andromeda, Edwin Hubble put a yardstick on the expanding universe (see page 30). Cosmology then became an observational science. Not only could astrophysicists test Einstein's theory of relativity, seemingly the most impractical part of all physics, but theorists began to need relativity in order to make sense of the observations.

Yet cosmologists face a continuing problem: because their objects of study lie vast distances from Earth, they are faint, and progress has depended critically on the development of giant telescopes and sensitive detectors.

BIG TELESCOPES

Time and again, the need for more light has driven astronomers to construct large telescopes. William Herschel (1738–1822) built several big reflectors, culminating in one with a mirror 48 inches (1.2 m) across. He cataloged many faint objects with it. The next large telescope was a 72 inch (1.8 m) reflector built by William Parsons, the Earl of Rosse, in 1845. The heavy tube of Parsons' telescope was slung between stone walls and could be moved only up and down. Working visually, observers used its powerful light grasp to explore distant celestial objects, discovering the spiral structure of the galaxy M51. But Parsons erected the telescope on his estate in Ireland and poor weather limited its usefulness.

HALE'S GREAT CONTRIBUTION

As the 19th century drew to a close, giant telescopes were beginning to drive the growth of astrophysics. During this period, George Ellery Hale (1868–1938) appears like an astronomy-impresario. Gifted with energy, organizational ability, and a flair for good public relations, Hale also had a touch of the poet. He found his true calling, not in solar physics where he did much valuable work, but in acquiring the instruments that made American astronomy foremost in the world.

Hale came from a wealthy family, and moved with ease among the industrial barons of the Gilded Age. In 1897 his first major project, paid for by a Chicago streetcar tycoon, was the

BELOW: For several decades, the 200 inch (5 m) Hale Telescope at Palomar Mountain was the world's largest telescope. Recent instruments surpass it in size but retro-fitting with new detectors has kept the 200 inch at the forefront. INSET: Observing with the Hale Telescope sometimes requires the astronomer to ride inside the instrument.

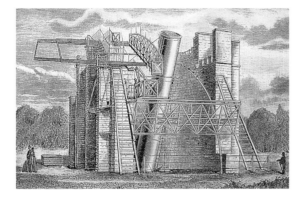

RIGHT: "The Leviathan of Parsonstown" was a 72 inch (1.8 m) reflector that cost William Parsons three years and £12,000 to build in 1845. Slung between two walls, it could follow an object for only an hour. It gave good views of galaxies, but fell into disuse because of bad weather and difficulties in handling.

Yerkes Observatory in Wisconsin, which still has the world's largest refractor telescope, a 40 inch (1 m). But Wisconsin skies are often cloudy and Hale turned to Mount Wilson, overlooking a then smog-free Los Angeles. The Mount Wilson Observatory was a joint project: Andrew Carnegie funded a 60 inch (1.5 m) reflector, and later John Hooker, a local merchant, paid for a 100 inch (2.5 m). Both telescopes produced outstanding research—Edwin Hubble first measured the size of the universe with the 100 inch.

But in time Los Angeles became a poor backdrop for deep-sky astronomy because of light pollution, and Hale set to work again. His goal was a 200 inch (5 m) telescope, to be sited on Palomar Mountain, in the dark-sky country between Los Angeles and San Diego. Work began in 1928, but World War II intervened and the telescope wasn't finished until 1948. Hale never saw it; he died in 1938, largely from overwork.

Hale's projects trace an important change in astronomy. Big telescopes are essential for cutting-edge research, but are so expensive that they demand national funding. After the 200 inch, large telescopes hit a plateau. The Soviet Union built a 236 inch (6 m) reflector, but, plagued with difficulties, it has never reached its potential. Telescopes couldn't grow until technological change brought new approaches.

COMPUTERS AND ACTIVE OPTICS

The advent of computer control means that telescopes can be built with smaller, cheaper, and simpler mountings. Computers also keep mirrors in shape, thus allowing much bigger ones to be built. This has led to large compact instruments such as the twin 400 inch (10 m) Keck telescopes on Hawaii's Mauna Kea—a size Hale could only dream about.

Another development is active optics, which deforms the incoming light to counteract the distorting effects of Earth's atmosphere. This method enables smaller telescopes to match the results of some of the earlier "glass giants."

RIGHT: One of the major themes in astronomy today is the search for origins— where do stars and planets come from? The photo, taken through the 150 inch (4 m) Anglo-Australian Telescope, shows the Rosette Nebula and open cluster NGC 2244 in Monoceros, the Unicorn. Here lies a cluster of young stars and the cloud of dust out of which they were born.

THE FULL SPECTRUM

The universe produces radiation over an enormous spectrum. This radiation travels in waves of various lengths—from the long radio waves, through the infrared and optical waves, and on into the shorter ultraviolet rays, X rays, and gamma rays. The only part of the spectrum that our eyes can see is the optical band, but our bodies can also detect long-wavelength infrared, which we call heat. Except for its wavelength, this infrared radiation is no different from blue, yellow, or red light.

Until the 20th century, astronomy was a visual science, crafting its view of the universe from what was visible at optical wavelengths. The emergence of radio after World War II gave astronomers a hint of the range of other phenomena, but it wasn't until they could build telescopes capable of detecting the full electromagnetic spectrum that astronomy began to explore the universe in all its variety.

LONGER WAVES

Radio astronomy (see pages 24–25) emerged as the first non-optical branch of the science, but it was soon followed by others.

On the short-wavelength side of radio lie the infrared and the microwave. Discovered in 1800 by William Herschel, infrared astronomy didn't blossom until the 1960s, when good detectors became available. Microwave astronomy followed a similar path. Both provide a way to detect radiation from the cosmic background (the fading glow of the Big Bang), clouds of interstellar gas and dust (out of which spring new stars and perhaps planetary systems), and shells and disks of dust around stars. The infrared is also the place to examine our own solar system's planets, especially Jupiter.

SHORT WAVES

Hopping over the optical region, the spectral region on the shorter side of the visible is the ultraviolet. Oxygen and ozone in Earth's atmosphere stop most ultraviolet radiation, so astronomers place observatories in orbit above the atmosphere (see pages 26–27). The ultraviolet region provides a look at processes in very hot stars, which evolve more quickly than cooler stars. The extreme conditions in hot

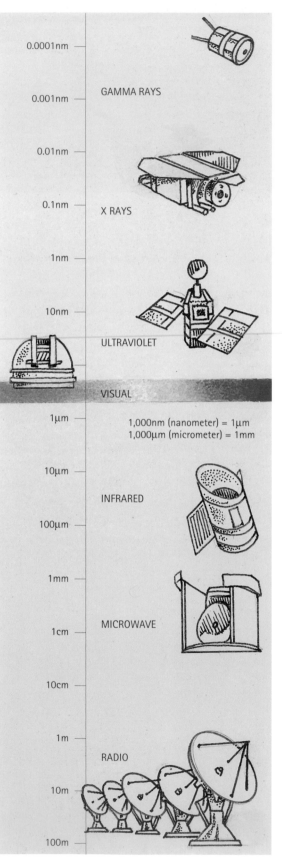

THE SPECTRUM

Our Sun produces light most strongly in one small part of the electromagnetic spectrum, which our eyes evolved to make use of. But the universe works with a broader palette of "colors." These range from high-energy gamma rays and X rays, through the visible, and down into the infrared, microwave, and radio parts of the spectrum. By developing instruments to detect this non-visual radiation, we sidestep the limitations of eyesight and begin to see the universe as it truly is.

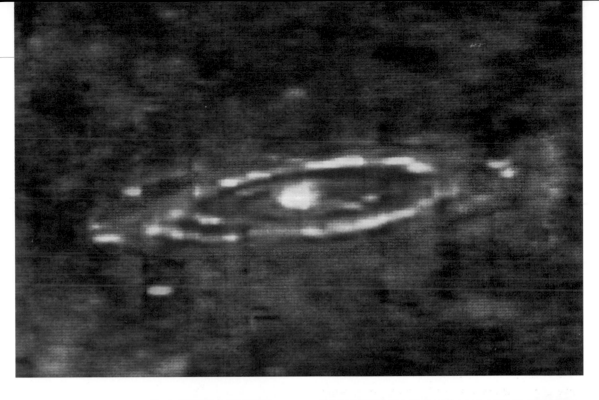

RIGHT: The nearest large galaxy to the Milky Way is M31, the Andromeda Galaxy. Its areas of warm dust and hot stars are shown in this map, made by an infrared satellite observatory. The galaxy's nucleus, seen as yellow, is bright because it contains many active stars and lots of dust. The galaxy's disk is cooler (redder) because it is less active, though hot knots of stars and dust appear here and there in its spiral arms.

RIGHT: The Andromeda Galaxy is rotating, with the lower right side approaching us, the other receding, in this radio image, which has been color-coded to map velocities. The irregular contours show that the galaxy is not spinning smoothly.

FAR RIGHT: At the high-energy end of the spectrum, the Andromeda Galaxy shows many hot spots of X-ray emission. The brightest region is the nucleus, a dense cluster of billions of large stars.

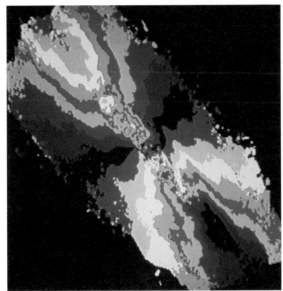

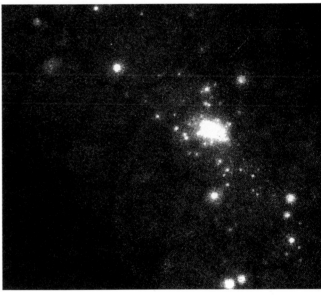

stars give astronomers better insights into how matter behaves. Even a cool star like the Sun has high-temperature regions, such as the corona (its outer atmosphere), where the ultraviolet view can be very informative.

X rays are shorter than the waves of ultraviolet radiation. Because X rays are blocked high in Earth's atmosphere, astronomers must use orbiting telescopes to study them. Instead of probing objects like a medical X-ray machine does, X-ray astronomy receives radiation from highly energetic processes that are occurring in the universe. Sources of X rays include objects hotter than about a million degrees. X rays are also emitted when high-energy charged

particles (such as electrons) interact with strong magnetic fields—a relatively common process in other galaxies and in parts of our own.

The shortest known radiation takes the form of gamma rays. Only the most energetic gamma rays reach the ground, so once again astronomers rely on orbiting detectors. Gamma rays are hard to deal with—they cannot be focused, and for the most part astronomers have to "see" them by the changes they induce in a detector. For all the difficulties, however, gamma rays can provide a look at some of the most exotic animals in the astrophysical zoo. These include black holes; the centers of active galaxies; and the energetic parts of the Sun.

RADIO ASTRONOMY

LEFT: Radio emission from the universe is weak, and radio telescopes need large collecting areas. The dish at Parkes, New South Wales, Australia, is 210 feet (64 m) across. (Note the technicians at left.) Radio telescopes can be linked together to resolve details finer than any of them are able to "see" alone.

The investigation of the universe at radio wavelengths was astronomy's first foray outside the realm of visible light. Heinrich Hertz had demonstrated radio waves in 1886, but the first detection of a cosmic source occurred only in 1932, when Karl Jansky, a Bell Telephone Laboratories engineer, accidentally discovered radio emission coming from the Milky Way. In the late 1930s, an American amateur astronomer, Grote Reber, constructed a steerable dish antenna and used it to map the Milky Way. Radio astronomy advanced little, however, until World War II.

INVESTIGATING RADIO EMISSIONS

In 1942 J. S. Hey in England realized that the odd signals radar sets were receiving came not from German jamming but from the Sun, then at maximum activity. Soon astronomers detected other radio sources, including a distant (but powerful) radio galaxy, dubbed Cygnus A.

A key point in understanding radio emission came in the early 1950s when astronomers

discovered that electrons spiraling in a strong magnetic field produce a characteristic radio signal. This turned out to be a common process in stars and galaxies, and it provided astronomers with a sensitive method for probing the magnetic fields of celestial objects.

Powerful sources of radio emission included Jupiter's magnetosphere, radio stars (many of them similar to the Sun), and hydrogen clouds in galaxies. It was at this time that astronomers used radio images of hydrogen gas in the Milky Way to map its spiral arms.

However, there were many strong radio sources that astronomers had difficulty identifying. In the early 1960s, radio astronomers worked out the position for a small but strong radio object. Optical astronomers then discovered a faint fuzzy bluish object in the same position and determined that it was a very distant galaxy. Soon such galaxies were termed quasi-stellar objects, or quasars (see page 51). Further research showed that quasars are the highly active nuclei of galaxies in which a massive black hole is broadcasting energy as it devours matter.

Another notable finding took place in 1967, when Anthony Hewish and Jocelyn Bell discovered a source of radio waves that pulsed with extreme regularity. They called it a pulsar (see pages 42–43). Soon they realised they had come across a rotating neutron star. Stars of this kind had long been predicted as the end-state of certain massive stars that explode as supernovas (see pages 40–41).

ABOVE: Quasar 3C 273 in Virgo is the nearest example of a quasi-stellar object. Seen here is a radio image of its jet, where red shows areas of highest emission as the jet smashes into a cloud of gas.

LEFT: The first radio telescope was this rotating aerial, built by Karl Jansky (foreground) to investigate static that was interfering with radio communications. The static turned out to be radio emission from the Milky Way Galaxy.

RADIO TELESCOPES

Because radio wavelengths are vastly longer than those of light, radio telescopes need to be huge to approach the same resolution as an optical telescope. Enormous dish antennas include the 250 foot (76 m) one at Jodrell Bank in England and the 1,000 foot (300 m) one at Arecibo in Puerto Rico. Radio interferometers are arrays of individual antennas that can be linked to produce the resolving power of a telescope many miles across. Examples of such arrays are Westerbork in the Netherlands, MERLIN in the United Kingdom, the Very Large Array (VLA) and Very Long Baseline Array (VLBA) in the United States, and the Australia Telescope.

Since the 1970s, radio astronomy has included microwaves. These fall between radio and long-wavelength infrared (see page 22), enabling astronomers to study interstellar gas and dust.

BELOW: The Very Large Array radio telescope consists of 27 movable antennas—each of them 82 feet (25 m) across—arranged in a Y-pattern in the New Mexico desert west of Socorro. Each arm of the Y is 13 miles (21 km) long. The antennas may be spread for maximum resolving power or clustered (as here) for greater sensitivity.

REFLECTED RADIO WAVES

An offshoot of radio astronomy is radar astronomy, in which radio signals are bounced off an object to create its image. Used as a radar dish, the Arecibo telescope (above) can map the shapes of asteroids and comets as far away as the main asteroid belt, and obtain detailed images of the surfaces of objects that pass near Earth.

SPACE TELESCOPES

If you've ever watched patterns of light and shade shifting across the bottom of a swimming pool, you can picture a major difficulty faced by astronomy. In much the same way that water distorts sunlight, Earth's atmosphere distorts and filters the cosmic radiation passing through it. To overcome this problem, astronomers some years ago began lifting telescopes above the atmosphere by placing them in rockets and satellites.

HUBBLE'S TRIUMPHS

Most famous of these orbiting observatories is the Hubble Space Telescope. Launched in 1990, repaired in 1993, and upgraded in 1997, it is the finest astronomical instrument ever built.

The Hubble Telescope's high, clear view has enabled astronomers to photograph the formation of galaxies when the universe was only a tenth its current age. And its discovery of variable stars in distant galaxies led astronomers

LEFT: The Hubble Space Telescope was named for American astronomer-cosmologist Edwin Hubble (see page 18) in reflection of the main intent of the telescope's builders, which was to probe to the edge of the universe—and to push that limit back still farther.

BELOW: The birth-cries of newborn stars are especially notable in ultraviolet light, visible only from above Earth's atmosphere. This image, taken by the ASTRO-1 telescope on board the space shuttle, shows in blue and white the brightest star-forming clouds as they trace spirals in the disk of M81, a galaxy in Ursa Major, the Great Bear.

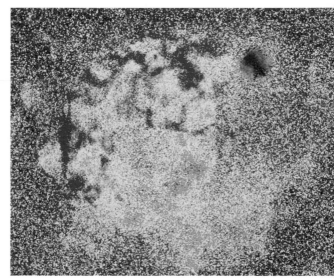

ABOVE: The center of the Milky Way Galaxy is relatively close to Earth, but only infrared radiation can penetrate the 26,000 light-years of dust lying in the way. In 1997 the Hubble Space Telescope took aim toward the galactic center and used its infrared camera to photograph the most massive star yet discovered: 100 times more massive than the Sun and 10 million times brighter.

ABOVE RIGHT: The remains of two exploded stars shine brightly in X rays, as seen by ROSAT. The Vela supernova exploded 12,000 years ago, and its expanding remnant makes a tangle of red and yellow streamers on the left. At the upper right, the 4,000-year-old Puppis A supernova remnant appears smaller and bluer, reflecting its higher energy and younger age.

to revise their estimate of the universe's size. Hubble has also explored the Orion and Lagoon nebulae where new stars are being born, and photographed pillars of dusty gas a light-year high inside the Eagle Nebula (see page 8). Closer to home, it has mapped the geology of the asteroid Vesta, monitored dust clouds on Mars, and tracked giant storms in Neptune's atmosphere.

NASA's Orbiting Observatories

But the Hubble Space Telescope was never meant to work alone. NASA designed it as one of four Earth-orbiting "Great Observatories," each covering part of the electromagnetic spectrum (see pages 22–23). The series is half complete. At the high-energy end is the Compton Gamma Ray Observatory, named for physicist Arthur Compton (1892–1962). Launched in 1991, the observatory discovered bursts of gamma rays coming from the most distant objects in the universe (see pages 52–53), and it found a huge cloud of antimatter (like ordinary matter, but with reversed electrical charge) lying 3,000 light-years above the center of the Milky Way Galaxy. Astronomers had never before found antimatter in such quantities.

Working at lower energies is the Advanced X-Ray Astrophysics Facility. This telescope is scheduled for launch in 1998, after which it will survey the entire sky at X-ray wavelengths for 5 to 10 years. Special targets will be hot

stars, supernova remnants, and the emission of X rays from distant clusters of galaxies.

The visible spectrum is covered by Hubble, so the last of the Great Observatories is the Space Infrared Telescope Facility, known as SIRTF. Due for launch in late 2001, SIRTF will study the clouds of gas and dust in which stars (and possible planetary systems) are born, and look for "brown dwarfs"—faintly luminous objects that are midway in size between planets and stars. It will also conduct deep-sky surveys, searching for evidence in the early universe of how galaxies began to form.

Other Satellite Telescopes

NASA has no monopoly on space observatories. German researchers launched ROSAT, the Roentgen Satellite, an X-ray survey telescope that made the first detection of X rays from a comet, when comet Hyakutake flew past Earth in 1996. In addition, researchers from many countries, including Russia, Great Britain, France, Italy, and Japan, have launched missions that focus on specific topics of astrophysical interest. A pioneering project was IRAS, the Infrared Astronomy Satellite, launched by NASA, Britain, and the Netherlands in 1983 to make the first infrared sky survey.

Most satellite telescopes work like those on the ground—astronomers from all over the world request observing time and are awarded it based on the scientific merit of their proposals.

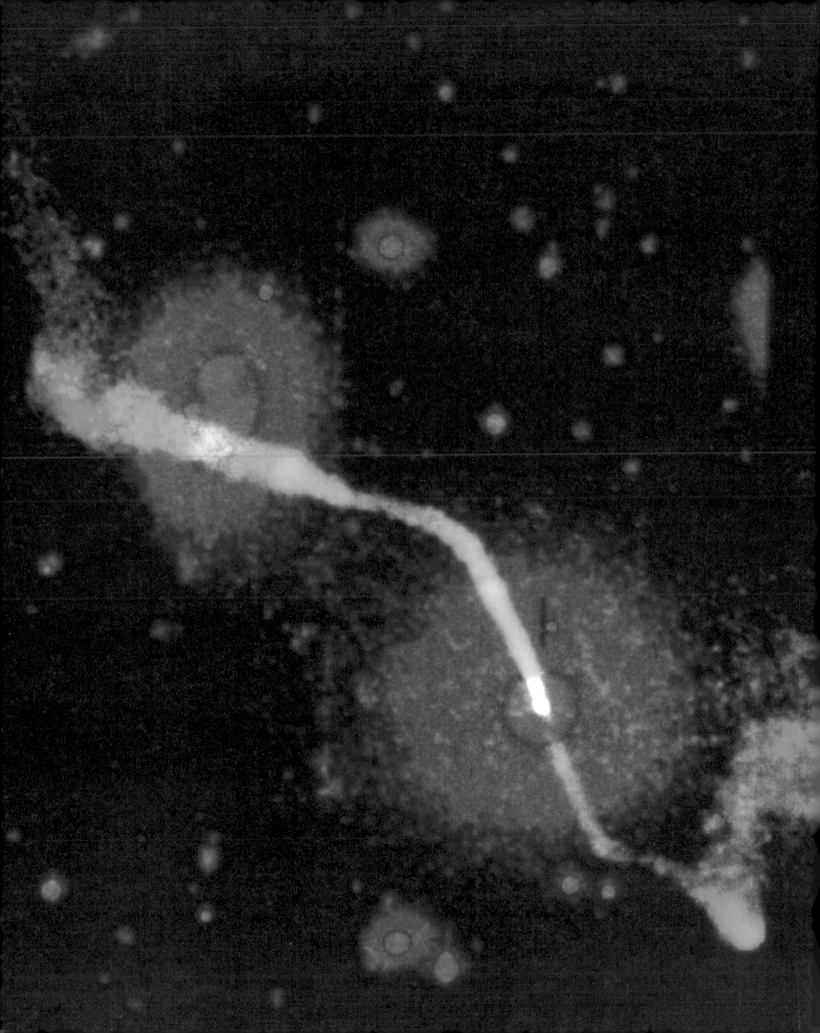

Understanding the Universe

Constantly expanding, the universe is of an unimaginable immensity. Our Sun is one of billions in the Milky Way, which is one of billions of galaxies in the universe. And stars and nebulae are only part of the picture: there are stars too small to shine, black holes that consume light, and perhaps even exotic invisible particles.

WHAT IS THE UNIVERSE?

It is certainly a challenge to picture the sheer enormity of the universe. The Moon is about 240,000 miles (385,000 km) away; the Sun, 93 million miles (150 million km); and the nearest star, Alpha Centauri, a staggering 25 million million miles (40 million million km). But the distance to Alpha Centauri is nothing compared with the vast reaches of the universe. The nearest galaxy to our own, Andromeda, is about a million times farther away. The most distant galaxies yet known are more than 10 million times farther still.

AT A STRETCH

For the very farthest objects, astronomers must abandon the notion of absolute distance and rely on the Doppler effect. This effect is more commonly manifested in the way the sound from an ambulance changes as it moves toward and away from you: on approach, the sound waves are squeezed to shorter wavelengths and the pitch seems higher, but as the vehicle passes the waves are stretched and the pitch drops.

The light from distant galaxies changes in much the same way. Because the universe is expanding, distant galaxies are all moving away from us, so their light is stretched—or made redder—as they recede. In the 1920s Edwin Hubble discovered that the farther away a galaxy is, the faster it is receding. So a higher "redshift" means that a galaxy is farther away. Eventually, astronomers will be able to translate redshifts into distances, but only when the value of the controversial "Hubble constant," which relates speed to distance, has been established.

Almost all the visible matter in the universe comes in the form of stars and nebulae. Stars are giant balls of gas that shine because of nuclear reactions in their cores (see page 34). Nebulae are huge clouds of gas and dust that are lit up by nearby stars (see page 38).

But astronomers have discovered that there is much more to the universe than meets the eye. The way stars move in galaxies shows that they are tugged by something invisible that has a huge gravitational pull. Most galaxies are probably surrounded by huge halos of this invisible "dark matter." In fact it could make up some 90 percent of the universe's mass.

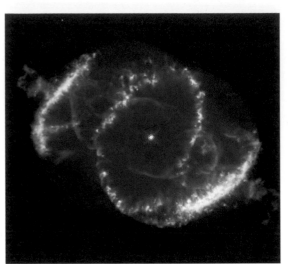

LEFT: *This striking image of NGC 6543, a planetary nebula in the constellation of Draco, was taken by the Hubble Space Telescope. The nebula formed when a star like the Sun exploded. The expanding gas shells are a rich "fossil" record of the dying star's late stages.*

What could this dark matter be? There are two main theories—MACHOS and WIMPS. MACHOS, or massive compact halo objects, are made of everyday matter but don't give off any light. Possible MACHOS are black holes (collapsed stars so dense they swallow light instead of emitting it, see page 42) and brown dwarfs (stars too small to shine). WIMPS, or weakly interacting massive particles, are more exotic. Smaller than atoms, they interact rarely with ordinary matter. If WIMPS really do make up the "missing mass," millions of them could be passing through your body every second.

So far, only a handful of MACHOS have been discovered and nobody has spotted a WIMP. Until astronomers solve this problem, though, we will be left in the dark about the bulk of the matter in the universe.

RIGHT: *In December 1995 the Hubble Space Telescope spent 10 precious days looking at an apparently empty patch of sky. This spectacular picture is the result of combining all the separate images. Some of the 2,000 or so galaxies that appeared could be the most distant ever seen.*

MEASURING THE UNIVERSE

Special distance measures are used to describe cosmic objects. For planets there is the astronomical unit (AU), which equals 93 million miles (150 million km)—the distance from Earth to the Sun. There are two measures for more distant objects. A light-year, the distance light travels in a year, is just under 6 million million miles (9.5 million million km). The other is a parsec, short for "parallax second." If you hold a pencil at arm's length and close first one eye and then the other, the pencil will seem to move. Knowing the distance between your eyes and how far the pencil moves tells you how far away it is. This principle is called parallax. A star that would appear to move one arcsecond if measured simultaneously from Earth and the Sun is said to be one parsec—about 3.3 light-years—away.

INFINITE OR FINITE?

Most scientists believe that the universe came into existence about 15 billion years ago in a cataclysmic explosion known as the Big Bang. At first there was nothing at all. Then out of nowhere came the first matter—a minute speck of seething exotic particles, hotter than 1000 billion degrees. This speck expanded and cooled until more familiar particles began to form—the neutrons, electrons, and protons that make up everyday matter.

Gradually the particles came together to form elements—mainly the lightest ones such as hydrogen and helium. The relative amounts of these elements in the universe today fit perfectly with the Big Bang idea. Eventually the elements collapsed under the influence of gravity to create galaxies, stars, and planets.

The Big Bang explosion produced a fireball of light that has been gradually cooling ever since, and what remains from the fireball now gently bathes the universe in microwave radiation—like the output from an enormous microwave oven—but at a temperature of just

−455°F (−270°C). This "cosmic microwave background" was discovered in 1965 by Arno Penzias and Robert Wilson, who were later awarded the Nobel Prize for physics. Along with the fact that the universe is expanding, the discovery of this background radiation is strong evidence for the Big Bang.

Astronomers don't understand why the universe began. Partly this is because these particles and light did not explode into an empty hall of space. Space itself began at the Big Bang, and so did time. So you can't ask "What came before?" because there was no before. The famous Cambridge astronomer Stephen Hawking once said that it would be like asking "What is north of the north pole?"

THE GREAT EXPANSION

The universe has been expanding from the time of the Big Bang, powered by the force of that initial explosion. But whether the expansion continues depends on exactly how much material—stars, galaxies, and dark

EVOLUTION OF THE UNIVERSE

1 **ZERO (15 BILLION YEARS AGO)** Universe is born out of nothing in a spectacular explosion called the Big Bang.

2 **A FEW BILLIONTHS OF A SECOND** Universe is a speck of exotic particles at incredibly hot temperatures.

3 **ONE MILLIONTH OF A SECOND** Universe has expanded considerably and contains particles—protons, neutrons, and electrons—that make up everyday matter.

4 **ONE MINUTE** Helium nuclei form from the protons and neutrons.

5 **HALF A MILLION YEARS** Atoms form. The intense radiation left over from the Big Bang begins to fade; universe darkens.

6 **SEVERAL BILLION YEARS** Huge clouds of gas start collapsing under effect of gravity, forming galaxies and stars.

7 **15 BILLION YEARS (THE PRESENT DAY)** Universe continues to expand. Galaxies jostle in vast clusters separated by empty reaches of space.

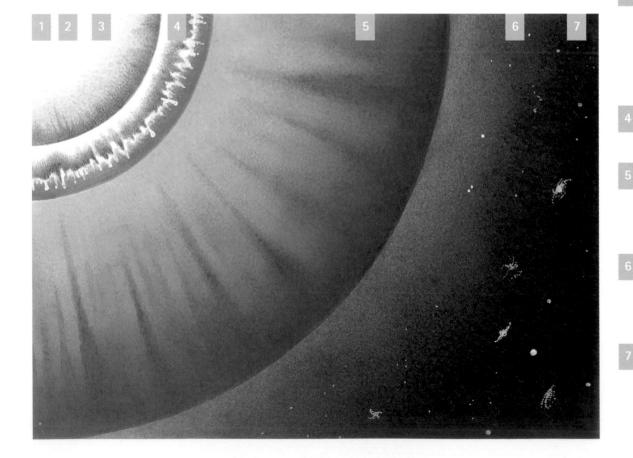

ABOVE: *The universe is bathed in microwave radiation, the afterglow of the Big Bang. This sky map produced by NASA's Cosmic Background Explorer satellite shows "ripples" in the radiation—some hotter than average (pink/red) and some cooler (blue). The cold spots are parts of the early universe that were relatively dense—these are the "seeds" from which galaxy clusters grew.*

matter—the universe has within it. Every bit of matter in the universe attracts every other with the force of gravity. If there is enough matter, the gravitational pull will eventually be strong enough to counteract the force of the Big Bang and stop the universe in its tracks.

The idea of this stopping point is captured in a number that astronomers call omega—the density of matter in the universe. If omega is much less than one, there is too little matter to

halt the universe and it will go on expanding forever. If omega is exactly one, the universe will eventually slide to a halt. But if it is greater than one, the universe will turn back on itself, and start contracting until it ends up in a cataclysmic implosion, a mirror of the Big Bang known as the Big Crunch. Since time has been flowing forward as the universe expands, some astronomers have wondered whether it would actually flow backward if the universe contracted, ending up back at zero at the Big Crunch.

Omega is extremely hard to measure. From all the light that is visible in the universe astronomers are able to estimate the amount of matter locked up in stars, and in the gas and dust illuminated by stars. But the universe also contains plenty of invisible dark matter—stars that are too small to shine, black holes that suck up light, and perhaps even exotic particles that no one has ever seen. At the moment, it looks as though omega is less than one, and the universe will continue to expand for ever. But astronomers are still looking hard for the extra mass that might change this picture.

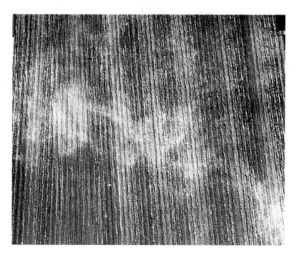

RIGHT: *This infrared image shows wispy clouds of interstellar dust and gas—probably hydrogen created in the fireball that followed the creation of the universe. Giant dust and gas clouds eventually condensed to form stars and galaxies.*

WHAT IS A STAR?

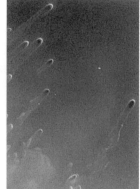

ABOVE: A detail of the Helix Nebula, taken by the Hubble Space Telescope.

LEFT: This dramatic image of the Helix Nebula shows the final days of a star like the Sun. The outer parts of the star exploded to create a glowing cloud known as a planetary nebula. The intersecting rings of the cloud span about 1.5 light-years.

When you look at a star, you're seeing a giant ball of fiery gas. Stars are born in a cloud of gas—made mainly of hydrogen but with a sprinkling of helium and sometimes other elements too. Something happens to make the cloud unstable—a nearby supernova explosion perhaps—and the cloud begins to collapse into a ball, creating an embryo star. As the star grows, its increasing gravitational pull drags more gas inward. When gas is squeezed it becomes hotter, so as the star grows the temperature inside rises. Eventually, the star becomes so big and dense that the temperature in its core reaches 18 million °F (10 million °C)—hot enough for nuclear reactions to take place—and it starts to shine.

MEMBERS OF THE MAIN SEQUENCE

All newborn stars start out by joining what astronomers call the "main sequence," which contains 90 percent of all stars. Main sequence stars generate the energy to shine by squeezing hydrogen nuclei together to create helium in a process called nuclear fusion. These stars come in a wide range of sizes, brightnesses,

and colors. The biggest and brightest are also the hottest. They can be up to 100 times the mass of our Sun, their surfaces are between 90,000°F and 13,500°F (50,000°C and 7500°C), and they shine blue or white. Next in the main sequence come smaller yellow stars such

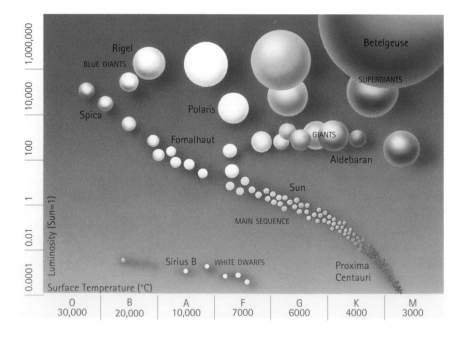

LIFE OF A STAR

1 **NEBULA** Stars begin life as a large cloud of gas and dust called a nebula.

2 **PROTOSTAR** Something triggers the cloud to collapse and it begins to form a denser ball called a protostar.

3 **MAIN SEQUENCE** When the star becomes dense and hot enough, it begins to burn fuel and shine.

4 **RED GIANT** When they run out of fuel, average-size stars swell up to become cool red giants.

5 **PLANETARY NEBULA** Eventually the giant's outer layers blast away to create a cloud called a planetary nebula.

6 **WHITE DWARF** The nebula dissipates and the core of the star is left behind as a cool, fading white dwarf.

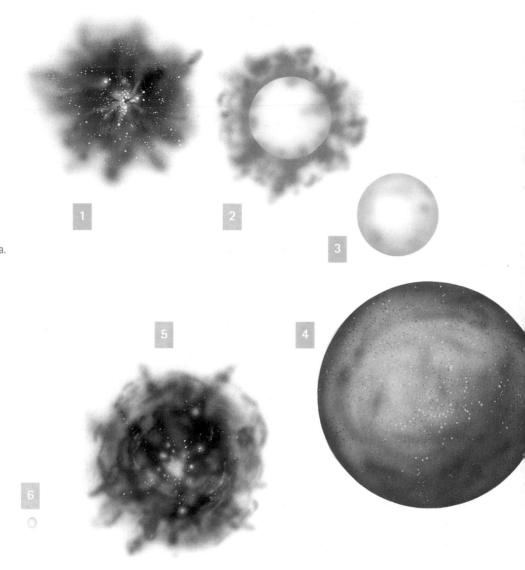

ABOVE: Betelgeuse is an orange-red supergiant star in the constellation of Orion. Cooler than the Sun, it is almost a thousand times bigger and will soon end its life in a supernova explosion.

HERTZSPRUNG–RUSSELL DIAGRAM

This compares star temperature and color. Brighter stars are usually bigger, and redder stars, cooler. Stars begin life on the main sequence (center), then evolve into larger, cooler stars (top right). Many then become faint white dwarfs (bottom).

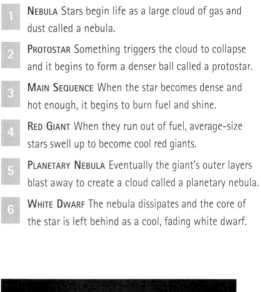

as the Sun. They are fainter, and have surfaces between 13,500°F and 9000°F (7500°C and 5000°C). Finally, there are the faint orange and red stars, which are small—as little as 8 percent of the Sun's mass—and cool, between 9000°F and 3500°F (5000°C and 2000°C).

Of the main sequence stars, the massive blue and white ones have the most fuel. But they also use it quickly—the biggest stars last only a few million years before their gas runs out. Smaller stars such as the Sun are more frugal and live for about 10 billion years, while tiny red dwarfs can last for thousands of billions of years.

GIANTS AND SUPERGIANTS

When stars run out of hydrogen in their core, they leave the main sequence and become giants or supergiants. Stars no more than eight times the mass of the Sun turn into giants. The central core is no longer producing energy, but the stars continue to burn hydrogen in a thin shell outside the core. The energy from this shell makes the star swell to a hundred times its original size. (When the Sun becomes a giant, in about 5 billion years, it will engulf Mercury, Venus, and Earth.) Because the stars still have the same mass, the outer layers have very low densities, and as they cool, the stars become red. All these giants will eventually turn into faint, cool white dwarfs scarcely bigger than Earth.

Stars more than eight times the Sun's mass become supergiants. Betelgeuse in the constellation of Orion is a famous red supergiant in the Milky Way. Eventually these stars explode in supernovas (see pages 40–41) and become neutron stars or black holes (see pages 42–43).

TYPES OF STAR

All stars change throughout their lifetime as they gradually evolve from birth to death, but some stars also alter on much shorter timescales. Stars known as intrinsic variables pulsate, changing their brightness and often their color in the process. One such variable, a red giant star called Mira in the constellation of Cetus, pulses every 331 days. The pulsations are dramatic: Mira varies in size from a maximum of 400 times the radius of the Sun to a minimum of just 200 times, and its temperature drops from 4000°F to 3000°F (2200°C to 1700°C).

USING STARS TO MEASURE DISTANCES

Other intrinsic variables are RR Lyrae stars and Cepheids. RR Lyrae stars are extremely old giant stars that formed soon after their galaxy came into existence. They typically change during periods of less than a day. Cepheids are very bright giant or supergiant stars with periods ranging from 1 to 70 days. Polaris, the North Star, is a Cepheid, pulsing once every four days or so.

Cepheids and RR Lyrae stars can both be used to measure cosmic distances. Cepheids pulse regularly and reliably, and ones that are intrinsically larger and brighter always pulse with longer periods. So if you measure the period, you know how bright the star actually is. Because of its distance, the star is dimmer than this as seen from Earth, so measuring how bright it appears in the sky tells you how far away it is.

The principle with RR Lyrae stars is similar, but as all stars of this description appear to have the same intrinsic brightness there is no need to measure the period. You simply need to know how bright the star appears to be in the sky. Other variable stars—called extrinsic, or eclipsing, variables—simply seem to change

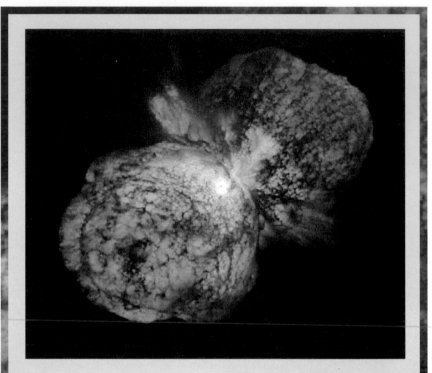

ETA CARINAE'S BRIGHTER TIMES

In the 17th century, Eta Carinae was visible to the naked eye but not especially bright. Subsequently it varied irregularly and by 1843 it had become the second brightest star in the sky. By 1868, however, it could no longer be seen. Astronomers now think a thick cloud of dust is swallowing its light. Eta Carinae lies 3,700 light-years from Earth and is intrinsically one of the Milky Way's brightest stars.

as viewed from Earth. These stars dim when a fainter companion passes in front of them and blocks some of the light, and then become brighter again as the companion moves away.

STELLAR COMPANIONS

Most stars have at least one companion star: as a singleton, our Sun is in the minority. Double stars probably form when several different parts of the parent cloud of dust and gas begin to

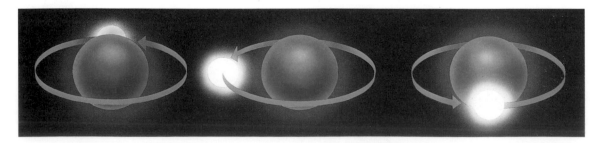

ECLIPSING VARIABLE
In some two-star systems, light from the brighter star (which is often the smaller one) can be dimmed when it passes behind its fainter companion.

collapse at once. This process does not produce only twins. The nearest star to Earth, Alpha Centauri, is a triplet star system. Two of the stars are bright, and orbit one another every 80 years. The third is tiny—if it were any smaller it would be unable to shine as a star at all. It weighs one-tenth as much as the Sun, and gives off 13,000 times less light.

Another prominent multiple star is Sirius, the Dog Star. The main star is a brilliant bluish supergiant, the sky's brightest star. But it has a faint white companion—a dead star known as a white dwarf, which no longer shines.

OPEN AND GLOBULAR CLUSTERS

Stars also gather in clusters. Open clusters such as the Pleiades can contain anything from about 20 to several hundred stars. They are loosely bound together by gravity, and perturbations from nearby nebulae will in time shake them loose. Such clusters rarely survive more than two circuits of their galaxy before falling apart, so they tend to be quite young—of the order of millions of years old. Globular clusters are denser and more tightly bound. They contain tens of thousands of stars and have probably existed since soon after their galaxy formed.

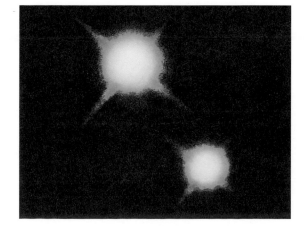

NEBULAE

Between their stars galaxies contain giant dense clouds of dust and gas—consisting mainly of hydrogen—which are called nebulae. The nebulae themselves give off no light, so on their own they would be impossible to see. But there are three kinds of nebula that stand out in the most spectacular detail against the background of space.

GLOWING COLORS AND DARK SHAPES

Perhaps the most beautiful of the three are the emission nebulae. These occur when a nearby hot star shines ultraviolet light on the hydrogen gas within the nebula. The light strips electrons from the hydrogen atoms. When the electrons rejoin their parents, they give off reddish light. Since many atoms in the nebula do this at the same time, the nebula is illuminated in glorious reds and purples. One of the brightest examples is the famous Orion Nebula, which is just visible to the naked eye as a fuzzy patch in the center of Orion's Sword (see page 127). It is lit up by four bright central stars known as the Trapezium cluster.

Another type is known as a reflection nebula. Nebulae of this sort shine because their dust reflects light from nearby bright stars. The small dust grains reflect blue light more efficiently than red light, so these nebulae are often blue.

The third type of nebula is dark. Dark nebulae contain the same mix of gas and dust as their bright cousins, but there are no stars nearby to illuminate them. However, we see dark nebulae when they block light coming from something behind them. Perhaps the most famous of these is the Horsehead Nebula, also in the constellation of Orion. Shaped like a horse's head, it stands out against the bright emission nebula that lies behind it.

STAR NURSERIES

With their dense mixtures of gas and dust, nebulae are perfect nurseries for creating stars (see pages 34–35). The Orion Nebula, at a distance of 1,500 light-years, is the nearest such nursery to Earth and the best studied. However, newborn stars tend to be shrouded in the dust of the surrounding nebula and are hard to see. In recent years many such regions have been spotted in the act of giving birth by the Hubble Space Telescope, and the images are providing astronomers with new clues about how stars, including our Sun, first came to life.

PLANETARY NEBULAE

Nebulae can also signal the death of a star. Close to the end of its lifetime, a star that is no more than eight times the mass of the Sun swells to become a red giant (see pages 34–35). This happens when it has all but exhausted its reservoir of hydrogen fuel.

Gradually the red giant blows off its outermost layers, which form an extended ring of gas. Left behind is a dense core, which weighs more than half as much as the Sun but is scarcely bigger than Earth. This burned-out core, known as a white dwarf, shines because of heat that is left over in its interior.

The light from the white dwarf illuminates the surrounding ring of gas, which is known rather misleadingly as a planetary nebula. In fact, the ring has nothing whatever to do with a planet—it was so-called because when 19th-century astronomers viewed it through a telescope, they thought it looked more like the disk of a planet than the point of a star.

ABOVE: *The Keyhole Nebula shown here is part of the Southern Hemisphere's Eta Carinae Nebula—the largest and brightest in the sky.*

RIGHT: *This hourglass-shaped nebula has intricate, colorful patterns on its walls, painted by glowing gas. At the core is a fading white dwarf star.*

BELOW: *Lying 1,600 light-years from Earth, the Horsehead Nebula is only visible because it obscures the light from the bright nebula behind it.*

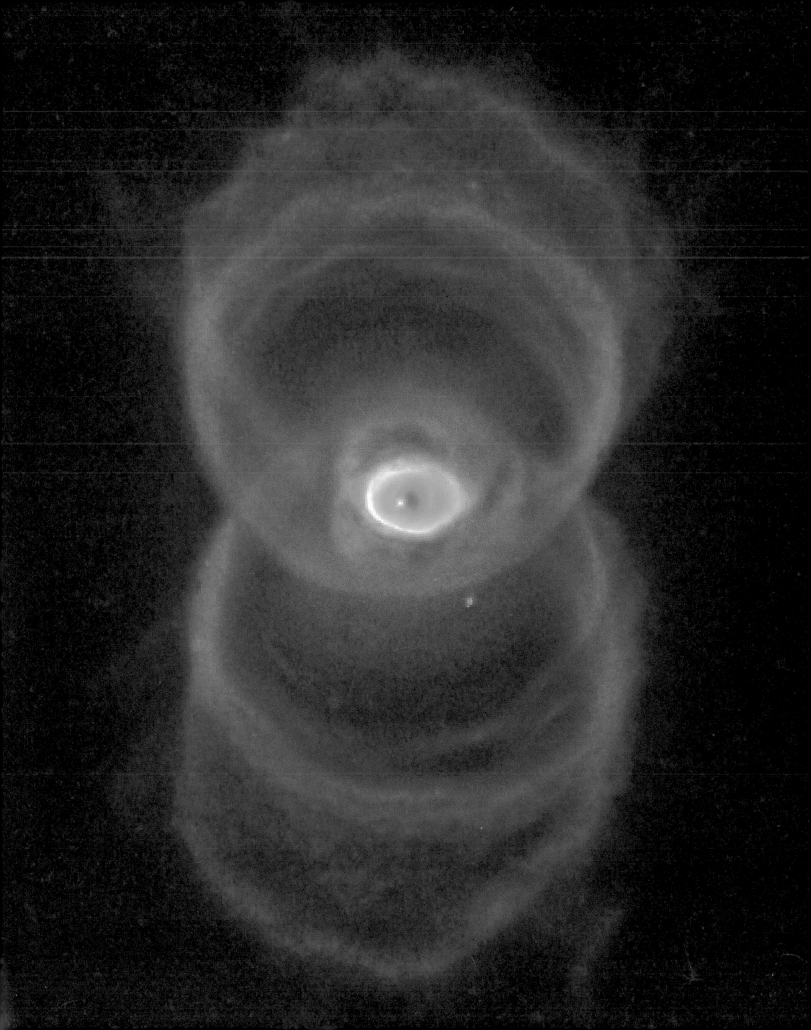

GOING SUPERNOVA

While some stars meet their end relatively gently, others go out with a bang. The bang is known as a supernova and is one of the most dramatic events in the universe. The most recent supernova in our own galaxy was spotted in 1604. No supernova since has appeared as bright as this one did, although Supernova 1987A—an exploded star detected in 1987 in the Large Magellanic Cloud—came close.

RUNNING ON EMPTY

There are two main types of supernova, which come from different types of star. The largest stars—those with at least eight times more mass than the Sun—start their lives as hot, bright, blue or white stars, burning hydrogen in their cores in nuclear reactions that produce helium and light. When a star this size runs out of hydrogen, after several million years, it swells up, cools down, and becomes a red supergiant, now burning helium in an attempt to cling on to life. Gradually, as it runs out of helium, it turns to heavier elements—carbon, oxygen, or neon. But the situation is becoming desperate for the star. It needs energy from nuclear reactions to balance the relentless pull it feels from its own gravity, and to stop it collapsing in on itself. Yet the heavier elements

produce much less energy than the hydrogen did. Eventually, the only remaining fuel source is silicon. But when that burns it creates iron, which does not burn at all.

With no means of producing energy, the star collapses, suddenly and dramatically. The inner part of the star becomes an incredibly dense core. The outer part falls inward, lands on the dense core, and explodes back outward at about 3,000 miles per hour (5,000 km/h). The star becomes one of the brightest objects in the sky, scattering its constituent elements into space.

ABOVE THE LIMIT

Supernovas can also come from white dwarf stars. White dwarfs are born when average-size stars turn into giants at the end of their lives, and then blow off their outer layers leaving a small white core. Most white dwarfs fade slowly into oblivion, but some are blessed with a companion star that helps make their demise much more dramatic. White dwarfs are stable only if their mass is less than 1.4 times the mass of the Sun—the so-called Chandrasekhar limit. If a white dwarf receives enough material from a companion star to take it above this limit, the result will be a set of runaway nuclear reactions and an almighty explosion, with stellar material tearing through space at about 7,000 miles per second (11,000 km/s).

When a massive star goes supernova, the dense core that it leaves behind can be a neutron star or a black hole, depending on the original mass of the star (see pages 42–43). Surrounding this dense central star is an expanding shell of gas called the supernova remnant, which can be visible for thousands of years. Exploding white dwarfs also leave remnants of gas behind them, but nothing is left of the original star.

Supernovas are the main way that elements are disseminated throughout galaxies. During a star's lifetime, its nuclear reactions create new elements. And more new elements are created in the high pressures and temperatures of the mantle of gas that is blown off during the explosion. It is these elements that make up everything you see around you, including your own body and all that you eat and drink.

LEFT: The Cygnus Loop is part of the expanding blastwave from a supernova that exploded 15,000 years ago. Astronomers study ancient supernovas such as this one to find out about the original blast.

RIGHT: In 1987 astronomers saw the brightest supernova in 400 years. This Hubble Space Telescope image shows the remnant in the center, ringed by ejected material. The blast probably blew a gas bubble, two segments of which are visible, lit by radiation.

BELOW: Supernova 1987A before and after it exploded. Though its parent galaxy—the Large Magellanic Cloud—is close by cosmic standards, it is still so far away that light from the explosion took 169,000 years to reach Earth.

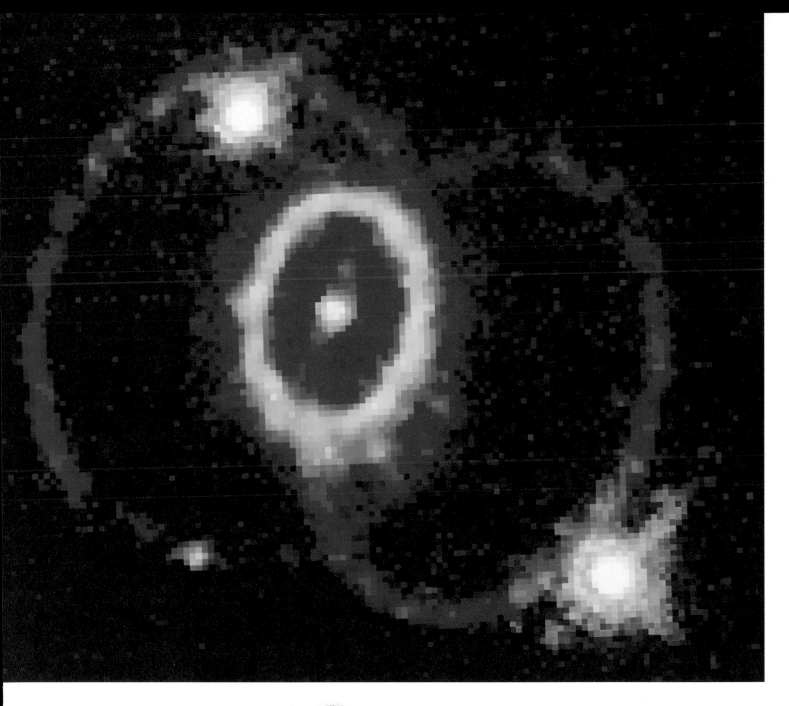

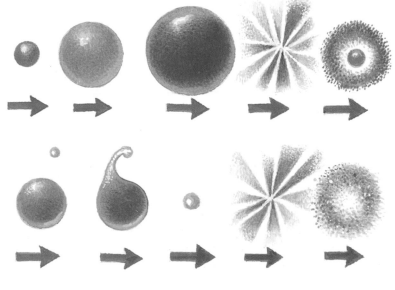

A Supernova Marking the Death of a Massive Star

A star at least eight times more massive than the Sun starts life as a hot blue or white star. As it starts to run out of fuel it swells and cools. Once its fuel supplies are exhausted, its core collapses and becomes incredibly dense. Its outer layers fall in on the core and explode in a supernova. The remaining dense core can be a neutron star or a black hole; the expanding shell of gas from the blast is called a supernova remnant.

A Supernova Occurring in a Binary System

In a binary system, a white dwarf sometimes drags material from its companion star, which is often a red giant. If the white dwarf's gain in mass is sufficient for it to become unstable, a series of nuclear reactions takes place and the white dwarf explodes as a supernova. All that remains following the explosion is the expanding shell of gas—the supernova remnant. The original star completely disappears.

PULSARS AND BLACK HOLES

Neutron stars are formed when a massive star collapses at the end of its life. Most of the star's material is blown off in a huge explosion known as a supernova (see pages 40–41). But the central ball of gas collapses in on itself until it becomes an extremely dense core made entirely of neutrons—tiny but heavy subatomic particles that are found in atomic nuclei. A neutron star is typically the size of a city and yet is heavier than the Sun. Just a teaspoonful of material from a neutron star would weigh as much as all the people on Earth put together.

PULSES OF RADIATION

When they are created, neutron stars spin extremely rapidly—once every second or so. Many of them send a pulse of radiation toward Earth every time they spin—much like a lighthouse beam. Astronomers believe that the Milky Way harbors more than a million of these "pulsars," although so far only a few hundred have been detected.

Because they are so dense, neutron stars have an intense gravitational pull. Many of these stars have an ordinary star as a companion, and material from the ordinary star is sucked

BELOW: Near the center of a distant galaxy called NGC 4261 lies this swirling disk, 800 light-years across. At its heart is a supermassive black hole, 1.2 billion times the weight of our Sun but hardly bigger than our solar system. The brightness at the center is probably a gas disk surrounding the black hole.

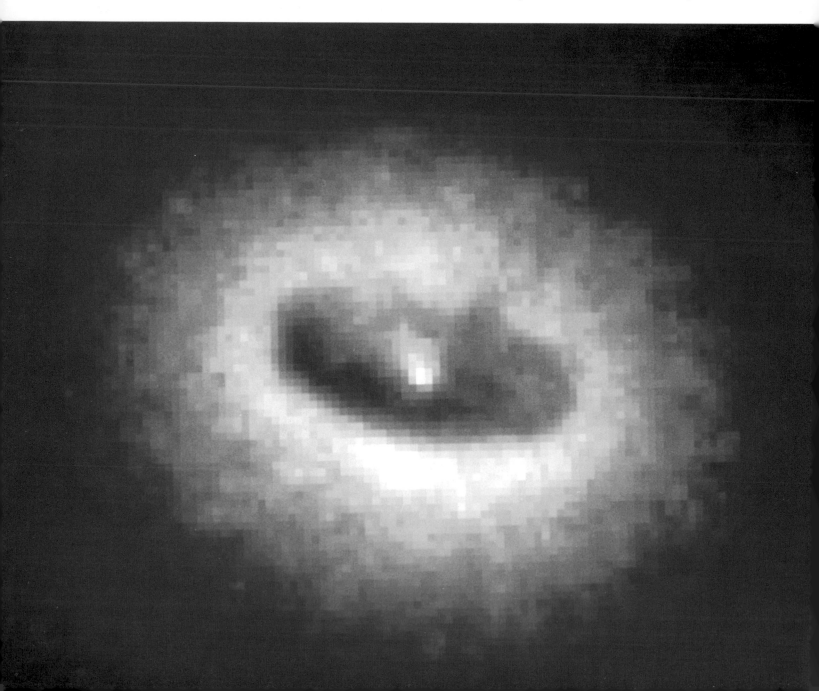

RIGHT: This is an X-ray image of an object called Cygnus X-1. Although it is invisible to human eyes, Cygnus X-1 is an intense source of X rays. These probably come from a hot disk of dust and gas surrounding a black hole about 10 times as heavy as the Sun.

BELOW: This galaxy, M87, is believed to contain a supermassive black hole. It spurts a jet of high-speed electrons thousands of light-years into space.

BELOW RIGHT: On July 4, 1054, Chinese astronomers recorded a new star. It was a supernova explosion, and this X-ray image shows the remaining cloud—the Crab Nebula. At left is the original star's core, now a dense pulsar.

in toward the neutron star, forming a disk of material that is gradually dragged onto the neutron star's surface. This process can "spin up" a neutron star, like whipping a top, until it rotates once every thousandth of a second.

PLACES OF INFINITE DENSITY

When a neutron star forms, it cannot be more than about twice the weight of the Sun. If it were any heavier than this it would continue collapsing under its own weight until it formed one of the strangest objects to be found in the universe: a black hole. Black holes are infinitely dense and they swallow up anything that comes close to them. Even light cannot escape their voracious gravity.

At the heart of every black hole, matter is crushed into an infinitely dense point known as a singularity, where the laws of physics cease

to operate. Surrounding the singularity is an imaginary surface called the "event horizon." This is the point of no return.

Spotting a black hole is very difficult, and astronomers usually try to find them by looking for their gravitational effects. There are several different kinds of black hole. The ones that form from supernovas are roughly the mass of a star. Others are "supermassive"— weighing in at up to a billion times the mass of the Sun—and lurk at the center of galaxies. These supermassive black holes could form when a large number of stars come close together in the dense core of the galaxy.

If you approached a black hole, strange things would begin to happen. Though time would seem to be passing at a perfectly normal rate, anyone watching you would see you slow down until you seemed to be in suspended animation. Space itself is so warped around a black hole that you would feel a much stronger gravitational pull on your feet than your head (assuming you were going in feet first), and this difference would rip you apart. Some scientists have speculated that if you could survive the tidal ripping and avoid the crushing force of the singularity at the black hole's heart, you might find yourself passing through into a completely different universe. Sadly, though, you could never return home to tell everyone about your adventures.

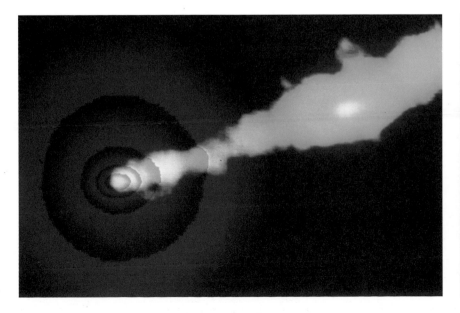

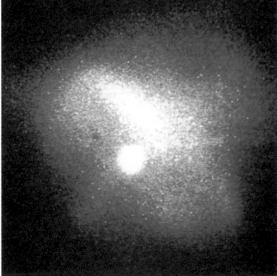

OTHER SOLAR SYSTEMS

Throughout history, the only planets that anyone knew about were Earth and its sisters orbiting the Sun. Then, in October 1995, a discovery sent shock waves around the world. For the very first time, a planet had been detected outside our solar system. The discovery opened the floodgates. Now there are nine new planets whose existence has been confirmed, and many more tentative findings.

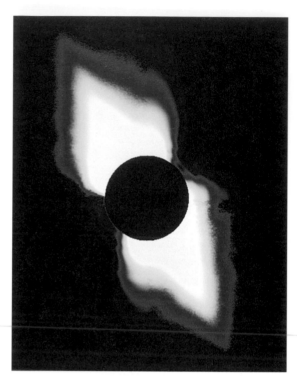

FINDING THE FIRST NEW PLANET

The first new planet to be detected is orbiting a star called 51 Pegasi in the constellation Pegasus. Such planets are so tiny and faint compared with their parent star that they are impossible to see directly, even with the most powerful telescopes. So astronomers use other methods in their hunts for solar systems like our own.

The trick used to find the 51 Peg planet is called Doppler shift (see page 30). An orbiting planet tugs slightly on its parent star, making the star move forward and backward as seen from Earth. This movement shows up as slight changes in the color of the starlight. It becomes bluer as the star moves toward Earth and redder as the star moves away. The size of the color change reveals the size of the planet, and the period over which it changes tells you the planet's orbital period.

Swiss astronomers used Doppler shift to work out that 51 Peg had a planet about half the size

LEFT: This color-enhanced dusty disk around the star Beta Pictoris is slightly warped, possibly because of the gravitational pull exerted by a planet or two hidden close to the star.

of Jupiter, orbiting it once every four days. An orbit this speedy was a major surprise. Jupiter, the biggest planet in our solar system (see pages 84–85), takes 12 years to travel round the Sun, and sits at a distance of about five times the distance of Earth from the Sun. But with an orbital period of just four days, the 51 Peg planet must nestle close to its star—a hundred times closer than Jupiter does to the Sun.

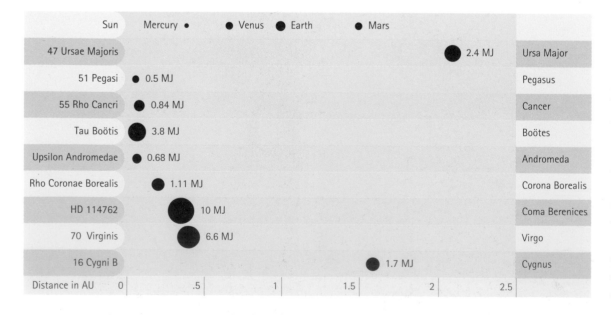

Sun	Mercury •	• Venus ● Earth	• Mars		
47 Ursae Majoris				● 2.4 MJ	Ursa Major
51 Pegasi	● 0.5 MJ				Pegasus
55 Rho Cancri	● 0.84 MJ				Cancer
Tau Boötis	● 3.8 MJ				Boötes
Upsilon Andromedae	● 0.68 MJ				Andromeda
Rho Coronae Borealis	● 1.11 MJ				Corona Borealis
HD 114762	● 10 MJ				Coma Berenices
70 Virginis	● 6.6 MJ				Virgo
16 Cygni B		● 1.7 MJ			Cygnus
Distance in AU 0	.5	1	1.5	2 2.5	

EXTRASOLAR PLANETS

In this chart, the nine new planets are compared with the inner planets of our solar system (not shown to scale). The figures beside each planet compare its mass with the mass of Jupiter (MJ). The new planets are much more massive than the solar system's inner planets (Earth is just 0.003 MJ), but orbit surprisingly close to their stars. The large planets in our solar system—the gas giants such as Jupiter—all orbit much farther out than any of the new planets do.

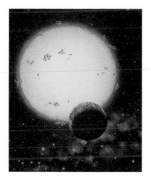

ABOVE: As shown in this artist's impression, the planet orbiting 51 Pegasi—like several other new planets—sits astonishingly close to its sun. It could not have been born there, but may have been dragged in later.

FORMATION THEORIES

The discovery of the 51 Peg planet caused havoc with ideas of how planets form. By looking at our own solar system, scientists had decided that big gassy planets such as Jupiter formed a long way from the Sun, and that only small rocky planets such as Earth could be born closer in (see pages 56–57).

Ironically, the 51 Peg planet's existence is now in doubt: though the jury is still out, one astronomer believes that the changes in the star's color come from the star itself pulsing in and out, and are nothing to do with an orbiting planet. But, in the meantime, several other planets have been found that show the same strange pattern—they all orbit much too close to their parent star for the old formation theories to hold. Astronomers now believe that something—perhaps interactions with the dusty disk from which the planets first formed—may have dragged the gas-giant planets close to their star. In the process, some smaller planets may even have been swallowed by the star.

If this new theory about planet formation is right, we are lucky that the same thing didn't happen in our own solar system, because Earth would almost certainly have been a casualty. Nobody knows why this didn't occur in our neighborhood, but one possibility is that the dusty disk didn't hang around for long enough after the planets had formed.

NOTHING LIKE EARTH

All the new planets are no more than spikes on a graph at this stage. No astronomer has seen what they look like, and nobody has found a planet anything like as small as Earth. But scientists believe that Earth-sized planets are the likeliest places for alien life to exist.

NASA and the European Space Agency are considering a range of proposals for telescopes that could detect Earth-sized planets. Most proposals involve combining data from several telescopes orbiting the Sun. The telescopes would be placed way out in the solar system, at about the same distance as Jupiter.

RIGHT: Taken by the Hubble Space Telescope, this image shows a young star surrounded by a disk of dust and gas. Eventually the material in this disk could clump together to create planets like those in our own solar system.

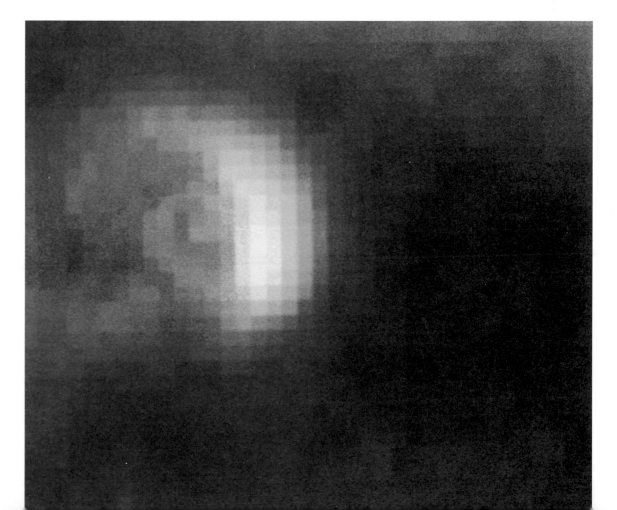

A UNIVERSE OF GALAXIES

Stars are not randomly scattered around the universe. Instead they gather in galaxies. A single galaxy can contain hundreds of millions of stars, as well as huge clouds of gas and dust. A few percent of the universe's galaxies have no particular shape and are known as "irregular," but the rest are either elliptical or spiral.

ELLIPSES AND SPIRALS

Seen from Earth, elliptical galaxies look like fuzzy, roundish areas of light. They have no internal structure to speak of, and mainly contain old, cool stars. There is little gas or dust between the stars. Spiral galaxies, on the other hand, look like lovely pinwheels in the sky. They are shaped like a flat disk with a central bulge, and have arms that spiral outward from the center. These arms are resplendent with bright young stars as well as the gas and dust from which the stars form.

Spiral galaxies usually have two arms, but some have been seen that have either one or three. Mighty Andromeda is the nearest spiral galaxy to the Milky Way. It is just visible to the naked eye as a faint patch of light in a constellation that is also called Andromeda.

LEFT: This misty, glowing patch is M32—a featureless elliptical galaxy. Roughly 60 percent of all galaxies are this shape.

In normal spiral galaxies, the arms emerge from the center of the galaxy—usually from opposite sides. In so-called barred spirals, however, the arms come from a bright bar that sits across the center.

Astronomers have tentative evidence that the central bulge in our own Milky Way is stretched out rather than spherical, suggesting that it could be a barred spiral.

WHY DO THEY VARY IN SHAPE?

Astronomers are still not sure why the different galaxies are the shapes they are. But because elliptical galaxies tend to contain only old stars

BELOW LEFT: The Virgo cluster of galaxies is the nearest of the major clusters, at just 60 million light-years away. With its vast gravitational pull, this collection of galaxies dominates our neighborhood, lying at the center of the Local Supercluster.

BELOW: The galaxy 4C41.17 was born in the very early years of the universe. One of the most distant of known galaxies, it lies roughly 14 billion light-years from us.

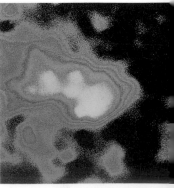

and very little gas, one theory is that they formed early in the history of the universe. They made their stars quickly, ran out of gas for new stars, and have been inactive ever since. According to this idea, spirals formed more gradually and are still using up their gas to make new stars. Some elliptical galaxies, however, may have been created when two spiral galaxies collided, destroying their pinwheel structure in the process.

FROM CLUSTERS TO SUPERCLUSTERS

Galaxies tend to gather in collections known as galaxy clusters. The Milky Way lies in a cluster of about 30 galaxies, known as the Local Group (see page 48). But many clusters are much larger. The nearest large one to us lies in the constellation of Virgo, near the north pole of the Milky Way. At its center, where the galaxies are most dense, the Virgo cluster contains more than 3,000 galaxies, all squeezed into a volume little greater than that of the Local Group. The

Virgo cluster itself lies at the heart of a much larger collection called the Local Supercluster, which also encompasses the Local Group.

The Virgo cluster contains mainly spiral galaxies, while the Coma cluster, which lies about six times farther away, has more ellipticals. Like many other clusters, the Coma cluster contains a huge cloud of gas at a temperature of about 200 million °F (100 million °C). The gas is probably heated as the galaxies stream through it. Astronomers are puzzling over why all these structures and superstructures formed, and this question—one of the hottest in the field—will probably not be answered for years.

THE MILKY WAY

The Milky Way Galaxy is vast—much larger and brighter than most other galaxies in the universe. It contains roughly a hundred billion stars, not to mention dense clouds of gas and dust where new stars come into existence at a rate of about ten every year.

A PINWHEEL IN SPACE

Most of the Milky Way's stars sit in a thin disk that orbits around the galactic center just as the planets in our solar system orbit around the Sun. The Sun sits within this disk, a little more than one-third of the way out from the center of the galaxy. If you look up on a dark night, you will see a bright band of white light splitting the sky in two like a giant highway. This band—a combination of the light from the vast numbers of stars within the disk—gave the Milky Way its name: to the ancient Romans the band looked like spilled milk.

At the center of the Milky Way lies a supermassive black hole weighing about a million times more than the Sun. Surrounding this is a roughly spherical bulge. Winding out from the center are spiral arms, which are embedded in the flat disk and house dense molecular clouds where new stars are born. If you could see the Milky Way from the outside, it would look like a beautiful pinwheel, its spiral arms illuminated by hot young stars.

Astronomers have managed to trace relatively nearby sections of the Milky Way's spiral arms, mainly by looking for young stars and associated emission nebulae (see pages 38–39). They have identified the Orion arm, which contains the Sun; the Sagittarius arm, which lies closer to the center; and the Perseus arm, which is farther out. More distant spiral sections are difficult to make out, because interstellar matter obscures them from our view. Astronomers are still uncertain whether the Milky Way has four spiral arms spiraling out from the galactic center or just two.

SURROUNDING THE GALAXY

The whole system is embedded in a sparsely populated spherical halo of old stars and star clusters. When the galaxy was first forming, it began as a spherical cloud, and only later collapsed into a flat disk. Many of the stars in the halo come from this very early time, and are almost as old as the galaxy.

The bright part of the Milky Way has a diameter of about 130,000 light-years. But extending outward for perhaps another 100,000 light-years is the mysterious dark halo, which gives off little light but contains most of the galaxy's mass. Astronomers know that the dark halo exists because of the gravitational pull it exerts on the stars in the galaxy. It could contain dark stars such as black holes (see pages 42–43) or it could be made up of exotic subatomic particles that are yet to be spotted.

The Milky Way contains every kind of star, from hot blue giants such as bright Sirius to cold red dwarfs such as faint Proxima Centauri—the closest star to the Sun at just 4.2 light-years away. Proxima Centauri is actually part of a system of three stars (see pages 36–37). Its two companions—Alpha Centauri A and B—are the next closest stars to the Sun, and together they form the third brightest star in the sky.

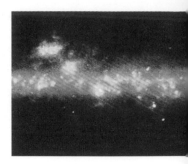

ABOVE: *Looking toward the galactic center, this infrared image of the plane of the Milky Way shows a pink band of radiation from dust warmed by starlight, and yellow areas of star formation. The galaxy's dusty disk hides nearby galaxies from our view.*

THE LOCAL GROUP

The gigantic Milky Way is not alone in its stately spin through the reaches of space. It is part of a cluster of more than 30 galaxies known as the Local Group. The group contains another giant galaxy, Andromeda (above), which lies 2.4 million light-years away. Together these two monarchs hold tight sway over their multigalactic empire: most of the remaining galaxies orbit around either the Milky Way or Andromeda in the way that planets orbit a star, and the rest are affected by the giants' huge gravitational pull.

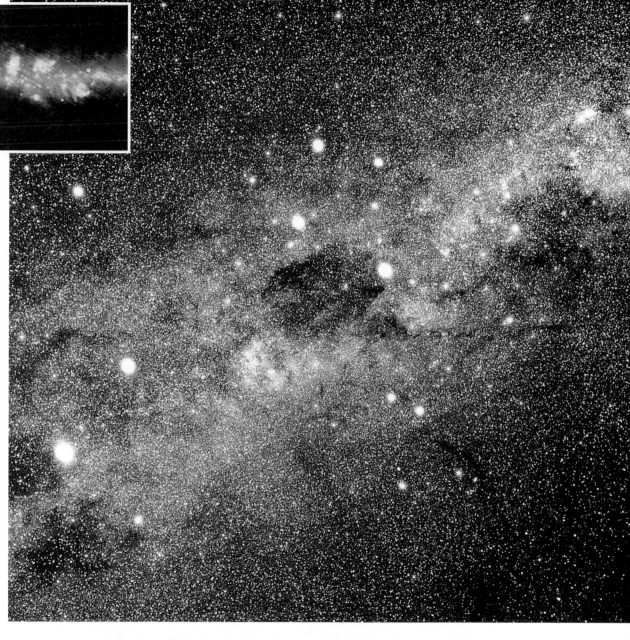

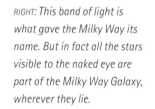

RIGHT: This band of light is what gave the Milky Way its name. But in fact all the stars visible to the naked eye are part of the Milky Way Galaxy, wherever they lie.

FORM OF THE MILKY WAY

Billions of stars form the disk of our galaxy. Seen from above (right), the Milky Way looks like a pinwheel and has either two or four spiral arms trailing through the disk. Seen side-on (far right), its plane is relatively thin with a bulge at the center.

galactic center

Sun

black hole

disk of galaxy

halo

central bulge

Sun

STRANGE GALAXIES

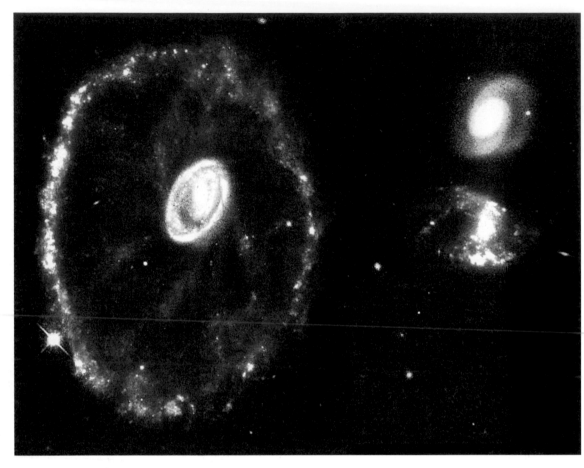

LEFT: *The Cartwheel Galaxy is the spectacular aftermath of an ancient collision. A small galaxy, perhaps one of those on the right, cannoned into the larger one, sending a wave of energy out into space like a ripple on a pond. This created several billion new stars in an expanding ring that is now large enough to contain our entire galaxy.*

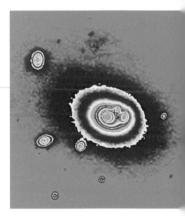

ABOVE: *NGC 6166 is one of the largest galaxies known. It is almost 2 million light-years across and is surrounded by a vast cloud of hot gas. As can be seen in this false-color image, at least five smaller galaxies lie in the same gas cloud, making the central monster a prime candidate as a celestial cannibal.*

I n the Southern Hemisphere constellation of Corvus, two galaxies called the Antennae galaxies are colliding head-on. The collision began about 500 million years ago, and the galaxies are named for the long stream of stars each is tearing from the other.

Astronomers believe that many galaxies have suffered similar collisions during the course of their lives. Some, such as the Antennae galaxies, bear the evidence in their strange shapes. For others the evidence is more subtle. Our own galaxy, the Milky Way, contains stars that appear alien, being the wrong age and composition for where they are positioned in the galaxy. These stars probably came from acts more like cannibalism than collision: the Milky Way recently swallowed up an entire dwarf galaxy in the constellation of Sagittarius, and some billions of years from today two other galaxies in the vicinity—the Large and Small Magellanic Clouds—will probably suffer the same rather grisly fate.

COLLISIONS AND CANNIBALISM

Collisions and mergers are tough on galaxies but relatively gentle on the stars they contain. The galaxies may be distorted before being ripped apart, but their stars are separated by so much space that they never collide. However, clouds of gas in the different galaxies can collide with spectacular fireworks, as the collisions trigger the birth of numerous bright new stars.

This phenomenon is thought to be taking place in many starburst galaxies—ones that contain enormous numbers of young stars that seem to have formed in a sudden massive burst. Tens of thousands of these galaxies have been discovered—some of them as far away as 1,000 million light-years—but astronomers are still not sure what triggers the bursts in all cases.

INTENSE ENERGY

Stranger still are active galaxies—ones that emit astonishing amounts of energy from what seems to be a tiny central source. The first of

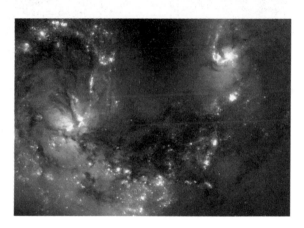

RIGHT: Hercules A is one of the brightest sources of radio waves in the sky. The waves come from two magnetic clouds, a million light-years across, supplied with energy by jets of electrons emerging from the galaxy's active core. Each of the bubbles blown by the jet on the right is bigger than the Milky Way.

RIGHT: When the Hubble Space Telescope snapped the Antennae galaxies in 1997, it revealed more than 1,000 star clusters bursting to life as the pair collides. At 60 million light-years from Earth, the Antennae are the nearest and youngest colliding galaxies known.

these to be discovered were Seyfert galaxies—giant spiral galaxies that are like the Milky Way, but with cores that are exceptionally bright. Imagine if the energy that is given off by our entire galaxy came from a region only a little bigger than our solar system. Roughly 10 percent of giant spiral galaxies have Seyfert nuclei, which is making astronomers wonder whether the Milky Way Galaxy was once as active as these galaxies and whether it might become active again.

Even more violent than active galaxies are radio galaxies—intense sources of radio waves that are many thousands of times "brighter" than normal galaxies. And brighter even than radio galaxies are quasars. These objects are at an almost unimaginable distance from Earth,

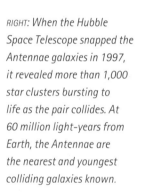

lying at the very edges of the universe. Also known as quasi-stellar objects, their light swamps that of their parent galaxies, making the quasars look like stars in the sky. Though quasars are not much larger than our solar system, they give out as much energy as a hundred galaxies.

SEARCHING FOR THE SOURCE

All these active galaxies produce astonishing amounts of energy from a very compact source. What could produce the energy? Most astronomers believe that the source is one of the strangest objects in the universe—a supermassive black hole. Lurking in the center of each of these strange galaxies is a black hole that weighs up to thousands of millions of times as much as the Sun and is gobbling up gas that flows in a disk around it. Astronomers have calculated that gas in a disk such as this can release huge amounts of energy before it finally plunges inward and meets its end.

THE LIGHT OF QUASARS

Quasars are extremely bright objects lying vast distances from Earth—the one at the center of this picture is about 1.4 billion light-years away. Astronomers use quasars as searchlights. From the edge of the universe, quasars send out beams that light up anything in their path. When the light reaches Earth it holds clues about intervening objects such as gas clouds or galaxies, and can even reveal where these objects lie between Earth and the quasar.

UNSOLVED MYSTERIES

Among the many remaining mysteries in the universe, two in particular have caught the attention of astronomers in the past few years, and have begun to yield their secrets. Both appear to come from explosions that are almost unimaginably energetic—perhaps the most dramatic in the universe.

BURSTING GAMMA RAYS

Roughly once a day, a brilliant flash lights up the sky and then disappears in a split second. We can't see these cosmic fireworks because they give out gamma rays—a kind of light whose wavelength is too short for our eyes to detect. Astronomers have known about gamma-ray bursts for nearly three decades but are still trying to work out where they come from.

Everyone agrees that gamma-ray bursts must be the leftovers of spectacular explosions. Debate centers on whether the explosions happen close by—in our own galaxy—or at the farthest edges of the universe.

The issue is difficult to resolve because it is hard to catch the bursts in the act. They are over so quickly that telescopes have no time to pinpoint them and study the surrounding area for clues. In 1997, however, a new Italian-Dutch satellite called BeppoSax located several bursts very quickly, giving a range of Earth-based telescopes the chance to scour the surrounding region. This scrutiny suggests that the bursts come from much farther away than our galaxy.

If this is right, the bursts are incredibly energetic, giving out as much energy in a few seconds as the Sun does in 10 billion years. No one quite knows what explosions could give off this much energy, but one candidate is a cataclysmic collision between two neutron stars that creates a black hole (see pages 42–43). There are plenty of even more exotic suggestions, though, and the jury is still out.

COSMIC RAIN

On October 15, 1991, an extraordinary object appeared in the sky. Though it was just a million million millionth of a yard in diameter, it packed the same punch as a tennis ball traveling at 180 miles per hour (300 km/h). It slammed into Earth's atmosphere, traveling within a whisker

LEFT: Studying the afterglow of gamma-ray bursts, such as this faint fuzz of visible light detected by the Hubble Space Telescope, will help astronomers to work out where these bursts originate.

of the speed of light. Then it disintegrated into a shower of particles, which were spotted on the ground by a detector in the Utah desert.

This visitor was a super-energetic cosmic ray. Usually nothing special, thousands of cosmic rays arrive at Earth every second, probably coming from supernovas (see pages 40–41). But the Utah particle was different. Though it was a simple proton—the center of a hydrogen atom—it had far more energy than a supernova can deliver. In fact, it had more than a hundred million times more energy than the Earth's most powerful particle accelerator can deliver. Small wonder that scientists called it the "Oh my God" particle.

Since then, scientists have picked up a steady stream of super-energetic cousins to the Utah particle, but there is no consensus on their source. Particle accelerators zoom their particles through strong electric fields. In the same way, the cosmic rays were probably whipped up to their high energies in a super-strong electric field somewhere in the cosmos. Electric fields exist wherever there are magnetic fields, and scientists know of many of these in the universe. But no known field is strong enough. A possible candidate is the field around a neutron star. The cosmic rays may even come from the same explosions that cause gamma-ray bursts. Over the next few years an army of detectors is being built that should help nail this cosmic mystery.

RIGHT: This cloud, in the southern constellation of Vela, is left over from a supernova explosion, which created a tiny but energetic neutron star (shown as a black dot in the inset radio image of the Vela supernova remnant). Neutron stars could be responsible for both gamma-ray bursts and super-energetic cosmic rays.

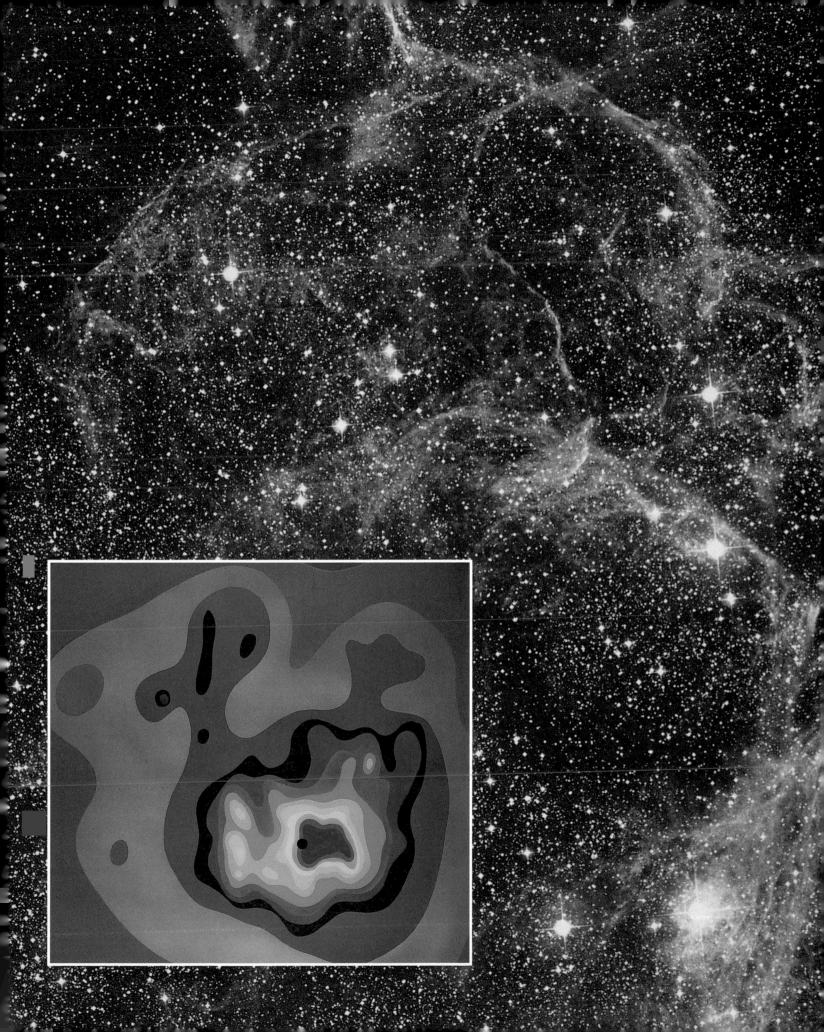

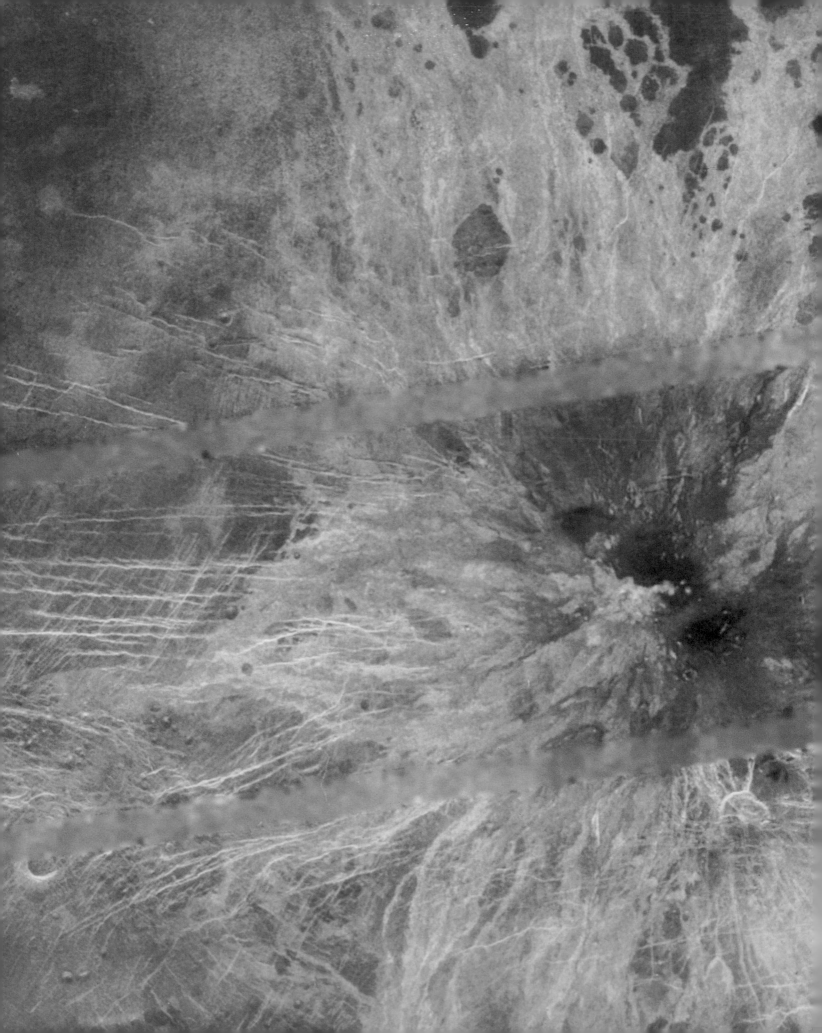

OUR SOLAR SYSTEM

THE SUN, A VAST GLOBE OF SEETHING GASES,
DOMINATES OUR SOLAR SYSTEM. AROUND IT
ORBIT THE NINE PLANETS AT VARIOUS
DISTANCES, MOST OF WHICH ARE ATTENDED
BY SATELLITES, SUCH AS EARTH'S MOON.
SMALLER BODIES—ASTEROIDS, COMETS, AND
METEORITES—ARE ALSO PART OF THIS
MYSTERIOUS AND BEAUTIFUL FRACTION
OF THE UNIVERSE.

THE BIRTH OF THE SOLAR SYSTEM

The Sun and the solar system were born in violence and chaos. But underlying the turmoil was an amazing degree of order visible today in the composition of the planets, moons, asteroids, meteorites, and comets around us.

FROM DUST CLOUD TO SUN

About 5 billion years ago, a large cloud of cold dust and gas was drifting around the center of the Milky Way Galaxy. It was average in size—if a beam of light could have punched through the smog from end to end, it would have taken a few hundred or a thousand years to do it. In places the cloud was thin and transparent—hardly denser than interstellar space. Other parts were more like a thick fog. It was slowly rotating relative to the stars around it.

The cloud might have remained that way forever if a star had not exploded nearby as a supernova. The explosion's shock wave pushed the denser portions of cloud enough that they began to collapse under their own gravity. The explosion also sprinkled the cloud with extra chemical elements.

The cloud fell in on itself, the densest part forming a core that grew larger and hotter as its gravity attracted material. The infall of debris sped up the cloud's rotation, just as whirling skaters spin faster when they pull in their arms. In a runaway process, the hot core developed into a protostar. Finally, the protostar grew large enough for the heat and pressure at its center to reach the flashpoint for thermonuclear fusion and the protostar became the infant Sun.

THE SOLAR NEBULA

But a star isn't all that was born from the cloud. As the dusty gas fell inward, perhaps 10 percent formed a broad disk around the protostar. In this disk, called the solar nebula, particles of dust collided and stuck together. Lightning from static discharges arced across the disk and through it, melting some grains and vaporizing others. Steadily the disk particles grew until they became planetesimals, bodies of rock and ice a few miles across. Collisions began to be more and more violent, no longer the gentle bump of particles meeting. New additions made their arrival with a bang.

While uncountable fragments were spawned, gradually the hordes of planetesimals accreted into a smaller number of larger bodies, the protoplanets. These in turn merged to form the planets. (Our Moon is the result of one such collision, caused when the early Earth was struck by a body at least one-third its own size. The Moon quickly coalesced out of the debris left circling Earth after the collision.)

Near the Sun, the disk was too hot for water to exist. The planets that formed there were made of tough elements such as nickel and iron. Farther out, beyond the "snow line" just inside Jupiter's present orbit, temperatures were cool enough for bodies to hold significant amounts of water, ice, and other elements easily destroyed by heat. Thus small rocky planets such as Earth lie near the Sun, larger gas-rich bodies such as Jupiter sit at a distance, and icy fragments called comets orbit on the outer fringes.

If its beginnings were slow, the formation of the solar system ended quickly. As the protostar became the Sun, it abruptly brightened, and its radiation swept away the remnants of the cloud that had fed the growing star and its family. The solar system was born.

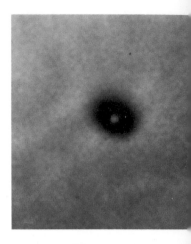

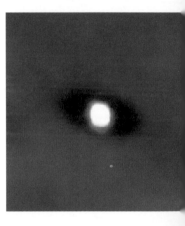

MAKING A SOLAR SYSTEM

Any theory of how the solar system formed has to explain a number of observed facts. For example, the planets all orbit in the same direction, matching the Sun's rotation. They have almost circular orbits lying in nearly the same plane. Also, the inner planets are rocky, while the outer ones are largely gas- and water-rich. This is how it probably happened.

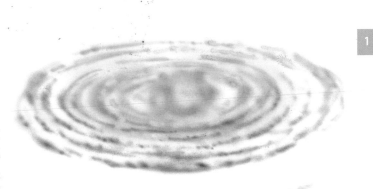

1 The cloud of dusty gas from which the Sun and solar system were born begins to rotate as it collapses, perhaps after a nudge from a nearby supernova explosion. Collisions among cloud particles force it to settle into a relatively thin disk rotating around the proto-Sun, all spinning in the same direction.

2 Lumps of dust in the disk collapse into solid bodies called planetesimals. Near the hot proto-Sun these are rocky in composition, while the planetesimals farther out can retain water, gases, and ice.

3 Planetesimals collide and form larger bodies termed protoplanets. As these grow by sweeping up the planetesimals in their vicinity, collisions become more violent.

ABOVE LEFT: Solar systems in the making 1,500 light-years from Earth, these smudges are clouds of dusty gas surrounding infant stars. The nebula at top is about five times the diameter of our solar system, while the lower one is ten times our solar system's size.

4 With protoplanets nearly at full size, gravitational encounters and collisions among them can tilt their rotation axes—or blow off material to make satellites, as happened with Earth and its Moon.

LEFT: The Great Nebula in Orion is home to more than 700 newborn and young star systems, including many that probably have planets. As astronomers study the conditions in the nebula they hope to learn more about how the Sun and planets formed.

5 Finally, after no more than a hundred million years, the newborn Sun's radiation blows away any material that has not been swept up by the planets—and the formation of the solar system is complete.

THE SUN, OUR NEAREST STAR

The Sun's everyday ordinariness tricks us into forgetting that we live less than 100 million miles (160 million km) from a raging thermonuclear reactor. That yellow-white disk drifting across the sky each sunny day is a ball of seething gases 865,000 miles (1.4 million km) across that weighs about 333,000 times more than Earth.

A LIFE-GIVING DWARF

The Sun is a dwarf star, like countless others among the 100 billion in the Milky Way Galaxy. It is 92 percent hydrogen, 7.8 percent helium, and all the other elements together add up to just 0.2 percent. Its structure appears to be quite simple, even if astrophysicists don't understand all the details.

At the Sun's heart is a gaseous core with a temperature of 27 million °F (15 million °C). In the core, hydrogen atoms bang into each other constantly. When four collide, they can fuse into one atom of helium and release a tiny

amount of energy. These energy sparks, released in prodigious numbers, power the Sun and thereby give us life. Every second, the Sun fuses about 600 million tons of hydrogen. And having done this for more than 4.5 billion years, the Sun is still only about halfway through its life.

MOVEMENTS OF ENERGY

But the energy doesn't reach us immediately. The tiny energy parcel spends 10 million years being absorbed and re-emitted before reaching the surface. Fusion occurs from the center of the Sun out to perhaps a quarter of its radius. Above this core lies a region in which radiation carries the energy. It reaches from the top of the core to 70 percent of the way to the surface. This is where the energy spends most of its time, trickling outward. The hot gas here is highly opaque—on average a parcel of energy flies only a fraction of an inch before being absorbed and emitted again.

On top of this layer the Sun's energy moves like boiling water. Heated from below, the gas rises to the surface, radiates energy into space, cools, then sinks again. This convective region forms the outer third of the Sun. Deep inside, the bubbles of hot gas are huge, but at the visible surface—the photosphere—they break down into granules about 500 miles (800 km) across.

The temperature of the photosphere is about 10,000°F (6000°C). It is only some 300 miles (500 km) deep, but displays many features. Best known are sunspots, small areas where the solar magnetic field stops hot gas from reaching the surface. Cooler than their surroundings, sunspots appear dark. For unknown reasons, their numbers rise and fall over an 11-year cycle.

THE SOLAR WIND AND AURORAS

A gale of charged solar particles—protons and electrons—constantly blows past Earth at supersonic speeds. When these particles interact with Earth's magnetic field, they spiral in and strike the atmosphere, ionizing the gas and causing it to glow in ghostly curtains, arcs, and rays. Called the aurora borealis or australis (depending on the hemisphere), these displays were once believed to be divine messages. Peak auroral activity takes place a year or two after sunspot maximum. The photo shows a Southern Hemisphere aurora seen from the space shuttle *Discovery* in 1991.

LEFT: Looking like flames, solar prominences are areas of cooler hydrogen gas caught in the high-temperature corona. Being dim compared with the Sun's bright photosphere, prominences can be seen only during total solar eclipses or when viewed (as here) at wavelengths that enhance their visibility.

RIGHT: A thin, extremely hot gas called the corona lies above the visible surface of the Sun. This image is from Yohkoh (Sunbeam), a Japanese satellite that has been monitoring the Sun's coronal activity at X-ray wavelengths since 1991.

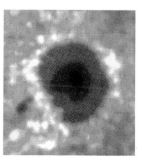

ABOVE: A sunspot the size of Earth shows its temperature in this infrared image. The coolest temperatures—about 8000°F (4500°C)—lie in the black core of the spot where magnetic fields have blocked the rise of hot gas from deeper within the Sun.

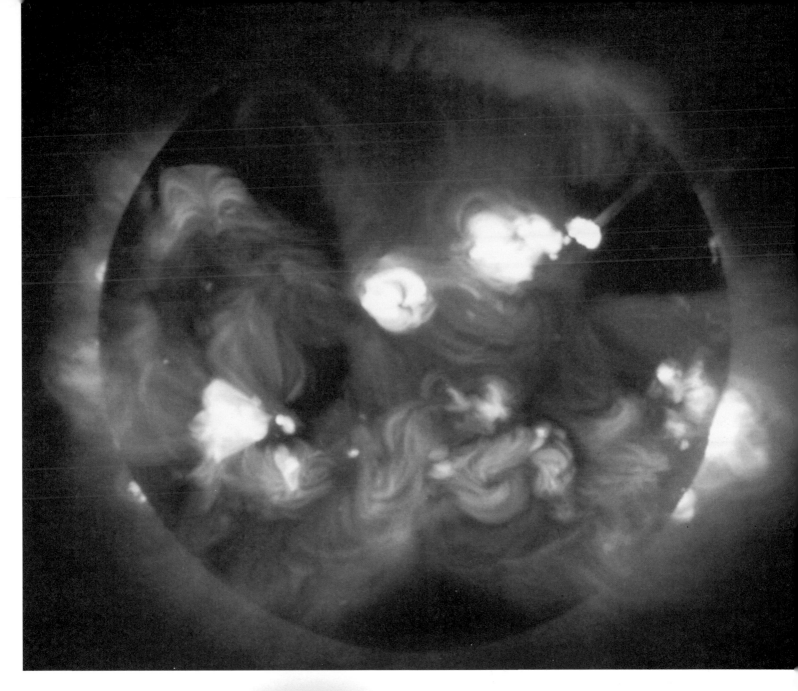

STRUCTURE OF THE SUN

Like all stars, the Sun is a machine for converting hydrogen and helium into heat and light. Now middle-aged, the Sun will continue to fuse hydrogen into helium for a few billions of years more. Then it will swell to become a red giant star, swallowing up Mercury, Venus, and possibly Earth in the process. Its ultimate fate is to throw off its outer layers in a shell of gas while the remnant of its core becomes a white dwarf star.

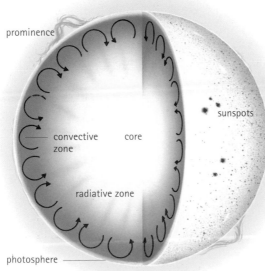

prominence

convective zone

core

radiative zone

sunspots

photosphere

SURFACE PHENOMENA

Prominences are another feature—huge loops of gas arching above the photosphere. Some are quiescent, others eruptive, but all move in response to the Sun's shifting magnetic field. More rarely, the Sun produces flares, powerful explosions that can lead to auroras on Earth and surges of current in electric power grids.

Above the photosphere lies a thin layer called the chromosphere, which is visible only when a total solar eclipse takes place. The chromosphere blends with the even thinner corona, also visible only during a solar eclipse, when it forms a pearly white halo around the Moon. Beyond the corona, a "solar wind" of protons and electrons blows into space.

THE PLANETS

From a stellar astrophysicist's point of view, the Sun *is* the solar system, and all the planets and moons around it are simply debris left over following its formation. After all, the Sun contains more than 99.99 percent of the solar system's mass. But within the remaining 0.01 percent we can discover much to explore: all the planets, moons, asteroids, comets, and meteorites. And since we consider one of those pieces of debris "home," our planet and its sibling worlds rightly draw intense interest.

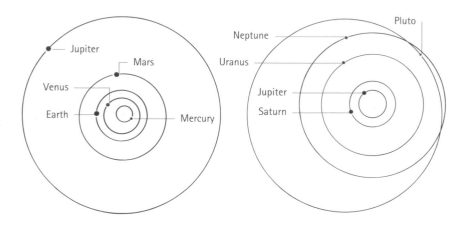

MEMBERS OF THE SOLAR SYSTEM

The family of planets falls naturally into two groups based on size, density, and chemical makeup. The first group is the terrestrial planets, consisting of Mercury, Venus, Earth, and Mars. These are small bodies made of rock, with densities three to five times that of water. Terrestrial planets have comparatively thin atmospheres, and Mercury has no atmosphere to speak of at all.

The second group of planets is the gas giants: Jupiter, Saturn, Uranus, and Neptune. These are all more than a dozen times more massive than Earth, and Jupiter outweighs our planet hundreds of times. While each has a tiny rocky core, it is buried under layers of hydrogen and helium thousands of miles deep. A gas-giant planet has a density near that of water, and Saturn's is actually lighter than ice.

The two categories omit Pluto, something of a special case. Pluto probably formed farther away from the Sun than it is now, and in time migrated inward. Little is known about this distant world—in size it is on the low end of the terrestrial-planet scale but its density is closer to that of the gas giants.

The solar system contains much more than planets, however. Between Mars and Jupiter lies the asteroid belt, a zone of small rocky debris prevented from coalescing into a planet by the gravity of Jupiter. Meteorites are fragments of asteroids, swept up by Earth after they were broken off their parent bodies by impacts. Moons? Every planet, except Mercury and Venus, has at least one moon, and each of the four gas giants has a ring system, of which Saturn's is by far the most

PLANETARY ORBITS
Inner planets lie close to the Sun, while those from Jupiter outward spread like growing ripples on the surface of a pond. Many planets may have formed at different distances from the Sun and migrated to their present orbits.

BELOW: There are places on Earth that resemble this Martian scene, photographed by the Pathfinder *lander. But each of the terrestrial worlds has its own geological style.*

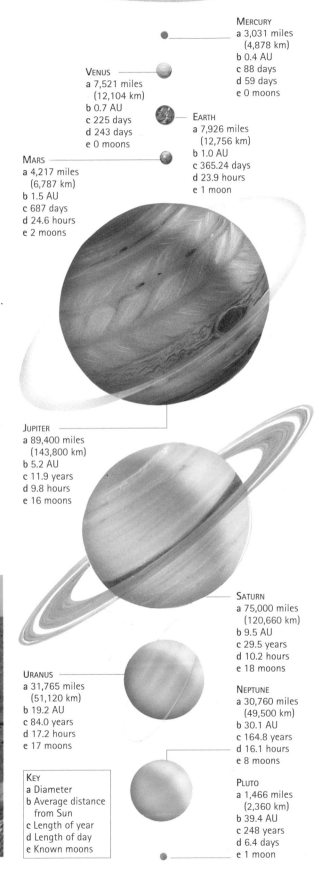

COMPARING THE PLANETS

Seen here to scale, with a portion of the Sun at the top, are the planets of the solar system. As spacecraft continue to explore them, scientists are learning that each planet is a world of its own, providing evidence about the solar system's origin and growth.

ABOVE: *Orbiting in a cold twilight far enough from the Sun to hang onto their supplies of volatile elements, the gas giants such as Jupiter retain a composition that generally matches the Sun's.*

extensive. Finally, on the fringes of the Sun's domain, halfway to the nearest stars, lies the realm of the comets—small icy bodies left in a kind of deep-freeze. These "dirty icebergs" have hardly evolved since the slam-bang days of planet formation, 4.6 billion years ago.

ORBITING THE SUN

Early astronomers sought to explain why all the planets orbited nearly in a plane, moving in the same direction around the Sun. Today this arrangement is understood to reflect the shape and structure of the solar nebula from which they formed (see pages 56–57). Their motions no longer mystify, either. Johann Kepler discovered in 1609 that planet orbits are ellipses, with the Sun standing at one focus. The planet with the most elliptical orbit is Pluto, followed by Mercury and Mars. The least elliptical orbit belongs to Venus, followed by Neptune and Earth.

Kepler also discovered that a simple relation connects the length of a planet's year with its distance from the Sun. The square of a planet's period in Earth years, he found, equals the cube of its average distance from the Sun in astronomical units. (Earth orbits the Sun at 1 astronomical unit, or AU.) Kepler's almost-mystical finding led directly to Isaac Newton's law of universal gravitation, one of the underpinnings of all modern physics.

MERCURY
a 3,031 miles
 (4,878 km)
b 0.4 AU
c 88 days
d 59 days
e 0 moons

VENUS
a 7,521 miles
 (12,104 km)
b 0.7 AU
c 225 days
d 243 days
e 0 moons

EARTH
a 7,926 miles
 (12,756 km)
b 1.0 AU
c 365.24 days
d 23.9 hours
e 1 moon

MARS
a 4,217 miles
 (6,787 km)
b 1.5 AU
c 687 days
d 24.6 hours
e 2 moons

JUPITER
a 89,400 miles
 (143,800 km)
b 5.2 AU
c 11.9 years
d 9.8 hours
e 16 moons

SATURN
a 75,000 miles
 (120,660 km)
b 9.5 AU
c 29.5 years
d 10.2 hours
e 18 moons

URANUS
a 31,765 miles
 (51,120 km)
b 19.2 AU
c 84.0 years
d 17.2 hours
e 17 moons

NEPTUNE
a 30,760 miles
 (49,500 km)
b 30.1 AU
c 164.8 years
d 16.1 hours
e 8 moons

KEY
a Diameter
b Average distance
 from Sun
c Length of year
d Length of day
e Known moons

PLUTO
a 1,466 miles
 (2,360 km)
b 39.4 AU
c 248 years
d 6.4 days
e 1 moon

MERCURY

The planet that orbits closest to the Sun, Mercury, is only 3,031 miles (4,878 km) in diameter. It is the smallest of the four inner planets—smaller than Jupiter's moon Ganymede or Saturn's moon Titan.

THE VIEW FROM HERE

Seen by the naked eye, Mercury is an elusive dot of light hovering near the horizon in the twilight. Ancient civilizations all knew of Mercury, even though it never wanders farther from the Sun than 28 degrees—roughly the width of an outspread hand held at arm's length. Often engulfed by the Sun's glare, the planet is difficult to see, and even some astronomers have never observed it with the naked eye.

In a telescope the view is not much better. Mercury looks pinkish gray and displays only the barest hints of faint surface features. Astronomers in the 19th and early 20th centuries thought that Mercury always kept one face turned toward the Sun. In 1965, however, scientists used radar to discover that Mercury spins on its axis every 59 days— exactly two-thirds of its 88-day year. Every two Mercurian years, therefore, the planet spins three times. This relationship means that Mercury always presents the same face to Earth whenever it is seen best.

THE WEATHER ON MERCURY

Cooked by a Sun shining nearly seven times stronger than at Earth, Mercury's noontime surface temperature reaches 800°F (430°C), while at night it drops to −280°F (−173°C). Ground temperatures vary wildly during the Mercurian day because there is no significant atmosphere to retain surface heat.

Oddly enough, there are some craters at Mercury's poles whose floors never feel the hot blast of sunlight. Radar results hint that these craters might contain thin deposits of water-ice under an insulating blanket of surface dust. Mercury formed at much too high a temperature to have had any water in its composition, but impacts by ice-rich comets may have given the planet a temporary atmosphere of water vapor, some of which could have condensed in the frigid polar shadows.

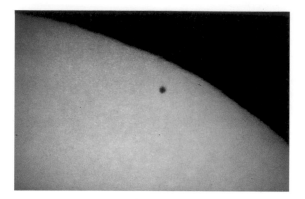

UP CLOSE

Mercury's only spacecraft visitor has been *Mariner 10,* which made three flyby passes in 1974 and 1975. *Mariner* photographed about half of Mercury and revealed a heavily cratered surface. The impact craters range in size from the smallest detectable, which is ½ mile (1 km) across, to the Caloris Basin, 830 miles (1,340 km) in diameter. This giant crater was created when a small asteroid struck early in Mercury's history.

While Mercury generally resembles the Moon, it does not have the big dark "seas" of lava found on the lunar surface (see page 70). Instead, its gently rolling lava plains appear more like the lunar highlands in composition, although they are much less heavily cratered. The plains feature many raised ridges, known as scarps, which tend to run north-south. Scientists attribute the scarps to some sort of compression force acting on Mercury's surface. They think the planet may have shrunk slightly as it cooled—even a 3 mile (5 km) contraction in diameter could explain the scarps.

BENEATH THE SURFACE

Mariner also found that Mercury has a nickel-iron center that fills about three-quarters of the planet's radius, proportionally much larger than any other planet's core. The large core may have formed that way, or it may be the result of Mercury losing some of its upper layers in a massive collision.

There are hints that Mercury's interior is still hot. The planet has a weak magnetic field, which is possibly generated by currents of molten nickel-iron flowing in its center.

LEFT: Mercury sometimes crosses the disk of the Sun, where a telescope prepared for safe solar viewing (see page 133) reveals it as a slowly moving black dot. These transits last several hours and happen on average 13 times a century. The next two will take place on November 15, 1999 and May 7, 2003.

RIGHT: Mercury's Caloris Basin is an impact crater big enough to reach from New York to Atlanta. The impact tapped lava sources deep inside Mercury, which filled the basin's interior (left half of image); the cracks and ridges formed as the lava cooled. Down the center of the image lies part of the basin's rim, now largely in ruins after innumerable meteorite impacts.

BELOW: *The Very Large Array radio telescope has mapped Mercury's temperature a few inches below the surface dust. In false-color red, the hottest spot shows a temperature of about 260°F (127°C).*

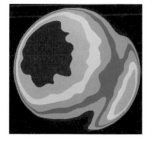

RIGHT: *At first glance, Mercury looks like the Moon. But even its most heavily cratered regions show fewer impacts than the Moon's southern highlands. This leads scientists to assign its surface a younger age, since older surfaces typically display more craters.*

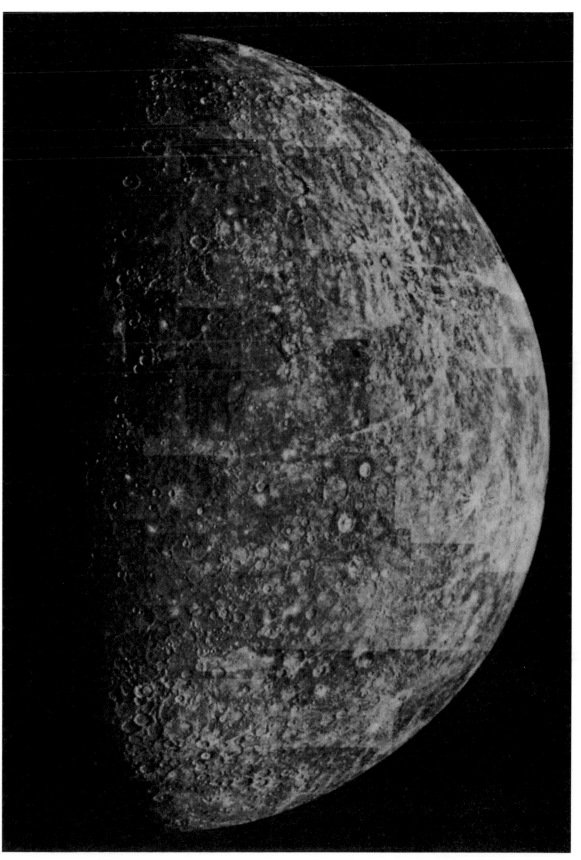

VENUS

If any planet could be described as hellish, it would be Venus. This world has all the searing heat of Mercury, yet it never cools off as Mercury does. Midnight on Venus is as hot as midday, and its surface rocks reach temperatures of 860°F (460°C), glowing dark red.

THE MORNING OR EVENING STAR

To the naked eye, Venus is the brightest object in the sky apart from the Sun and Moon—it can even cast a faint shadow. Viewed in the twilight of evening or morning, this beautiful "star" has often been mistaken for a UFO or an airplane with landing lights on. Earlier civilizations saw Venus as a god, and the Maya of Central America built a complex calendar around its celestial movements.

Like the Moon (see page 71), Venus goes through a cycle of phases. A telescope can reveal these, but will show few other features. The planet is completely shrouded in thick clouds made from droplets of sulfuric acid. They form in the planet's dense carbon dioxide atmosphere, some 90 times heavier than Earth's atmosphere. The carbon dioxide creates a runaway greenhouse effect that is responsible for the scorching surface temperatures. Sunlight, which is twice as strong at Venus as at Earth, penetrates the clouds and heats the surface. This heat radiates back at infrared wavelengths (see page 22), but because the carbon dioxide in the atmosphere blocks infrared, the heat is locked in.

TIME AND MOTION

For centuries, planetary astronomers lacked a view of Venus' surface and could do little besides tabulate the planet's size and motions. It has a diameter of 7,521 miles (12,104 km), about 5 percent smaller than Earth's, and orbits the Sun every 225 days.

On infrequent occasions—the next two are June 8, 2004 and June 6, 2012—Venus appears to cross the face of the Sun as seen from Earth. Such transits once helped astronomers measure the scale of the solar system, but radar has made the technique obsolete.

In 1962 radar enabled scientists to measure Venus' 243-day spin, which surprisingly goes backward compared with most other planets. This produces strange effects. With a Venusian "day" being longer than its year, if you were on Venus, you'd see the Sun rise in the west, cross the sky, and set in the east some 59 Earth days later! Early radar scans also detected a few surface features, which generations of spacecraft have since explored in much greater detail.

SPACECRAFT VISITS

The only photos taken on the surface cover just a few square yards and show gravel and sand mixed with slabs of lava. They were made by the Soviet *Venera* landers, which quickly succumbed to the intense heat and pressure. The American *Pioneer* and *Magellan* probes surveyed all of Venus from orbit using radar to create a global picture. They discovered a planet-wide museum of features—buckled and folded mountain ranges, enormous volcanoes, extensive lava flows, sinuous lava channels, and chunky lava domes. (See also page 146.)

Impact craters, however, were rare—fewer than 1,000 in total. Their scarcity indicates that Venus' surface is young in geological terms, perhaps only 500 million years old. While this is a long time by human reckoning, compared to the age of Venus—about 4.6 billion years—it qualifies as recent. Scientists think the planet must have experienced a global volcanic cataclysm that abruptly erased the traces of earlier eras. They are now trying to determine what geological forces might drive such an upheaval and whether Venus is still volcanically active.

ABOVE: A crescent Moon about to pass in front of Venus makes a striking sight. Venus glitters brightly in our morning or evening skies, and is often referred to as the morning star or evening star.

LEFT: Although blank to the eye, Venus' clouds display features when seen in ultraviolet light. From these, scientists learned that the upper atmosphere rotates every 4 or 5 days, far swifter than the planet's 243-day spin and much faster than the sluggish winds at the surface, which blow at only a few miles per hour.

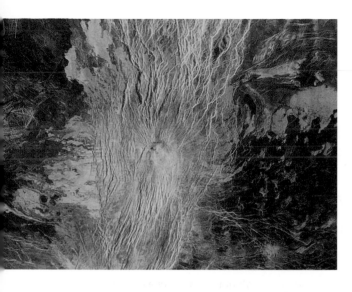

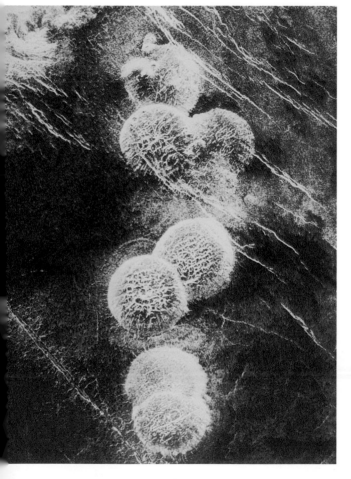

ABOVE: *Stiff lava builds pancake domes such as these in Venus' Alpha Regio. Each dome is about 2,500 feet (750 m) high and 12 miles (20 km) across.*

TOP: *Venus' rift zones, such as Devana Chasma in Beta Regio, resemble rifts on Earth. These surface features stem from tension in the planet's crust.*

ABOVE: *This false-color computer perspective shows two Hawaiian-type shield volcanoes: Sapas Mons (foreground) and Venus' tallest volcano, Maat Mons (background).*

TOP: *Mountain belts of Aphrodite Terra write a tangled scrawl across Venus' equator in this false-color view made from Magellan images. Venus' activity has given it features unseen on Earth, such as the oval rings in the lower left of the disk.*

EARTH

Perhaps the most valuable gift of the space age has been the knowledge that Earth really is a planet. This was hardly a secret, but it wasn't until scientists had more than token information about the other planets that they began to see how Earth fits into the solar system.

THE WATERY PLANET

Our planet is the largest of the terrestrial group. It has a diameter of 7,926 miles (12,756 km) and orbits the Sun at 1.0 astronomical unit, or 93 million miles (150 million km). Our year lasts 365.24 days. Earth's atmosphere is rich in nitrogen and oxygen, unlike the carbon-dioxide atmospheres of Venus and Mars. Ours is the only planet where abundant amounts of water exist in solid, liquid, and vapor form. And Earth also appears to be the only place in today's solar system where life exists.

Internally, Earth consists of three parts: core, mantle, and crust. While its center lies only a few thousand miles beneath our feet, it is wholly inaccessible. Still, scientists have deduced its properties by studying the ways in which seismic waves from earthquakes move through the interior of the planet. These show a partly molten core of nickel-iron that

is about 4,400 miles (7,000 km) in diameter. Scientists think that slow movements in the molten part of the core create currents that generate Earth's magnetic field.

Earth's mantle contains high-density basaltic rocks, rich in iron and magnesium. It extends from the top of the core almost to the surface. The mantle's top layers lie beneath the crust—the part we live upon. The thickness of the crust varies, being 5 or 6 miles (10 km) thick in the ocean basins, but reaching a depth of 50 or 60 miles (100 km) under the continents. The continents are made of granitic rocks, and float like rafts on the heavier crust beneath.

LEFT: Life on Earth did not arise in an orderly way: mass extinctions from time to time created opportunities for some forms of life while dooming others. We stand today—along with the other creatures that share Earth with us—at the end of a random chain of geological and biological accidents.

BELOW: The Twelve Apostles tower along the southern coast of Australia. They are made of layers of limestone that formed as seabed millions of years ago. In a cyclical process as old as the planet itself, waves erode the coast, returning the minerals to the sea.

THE STRUCTURE OF EARTH

Moving in cycles that last millions of years, warm currents in the mantle create new crust in rift zones on the continents and under the oceans—part of the process of plate tectonics. Scientists think that sea water is a key ingredient in plate tectonics, easing the subduction of one crustal plate as it collides with another.

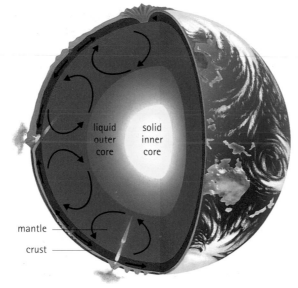

liquid outer core

solid inner core

mantle

crust

GEOLOGICAL ACTIVITY

Unless one lives in the neighborhood of an active volcano or a fault, the processes of geological change seem unreal. Yet our world is quite active by planetary standards. The reason lies in Earth's abundant supply of internal heat, left over from the energy of its formation and also created by the decay of radioactive elements such as thorium and potassium that occur naturally in rocks.

Earth's crust is broken into a dozen or more tectonic plates that collide and interact, riding on the back of the warm, plastic rock in the upper mantle. Where plates spread apart, such as the Mid-Atlantic Ridge in the ocean, new crust forms about as fast as fingernails grow. Where plates collide they can raise great mountain ranges such as the Himalayas, formed where the Indian Plate is pushing under the edge of the Eurasian Plate. Sharply defined, stationary plumes of hot rock shoot up through the mantle, creating chains of volcanoes as the mobile crust passes over the plume. This is the way the Hawaiian Islands formed.

WIND AND RAIN

The other partners in sculpting the world around us are the atmosphere and water in all its forms. Wind and rain work slowly but steadily, removing material from the land and returning it to the sea. By attacking the softest rocks and soil first, erosion shapes the landscape.

The air and the seas also serve to moderate Earth's climate, carrying heat into arctic regions and bringing cool currents to the tropics. The hottest deserts and coldest ice caps are far more habitable than they would be otherwise.

RIGHT: If water is the agent of geological destruction, then fire is often its creative sibling. A narrow column of molten rock shooting up from Earth's mantle reaches the surface under the Big Island of Hawaii, where one of the vents is the volcano Kilauea.

EARTH AND THE SUN

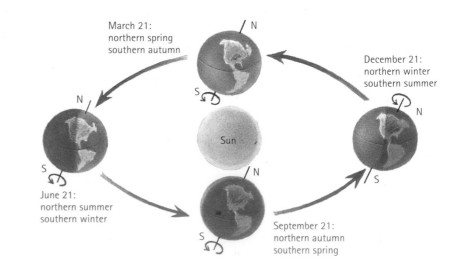

LEFT: People of the middle latitudes in both hemispheres are used to four seasons, but tropical regions such as Rajasthan tend to experience just a dry season and a wet. In the northern tropics, cool dry winds blow mainly from the northeast during northern winter. During the northern summer, warm wet winds from the southwest often bring torrential rains.

In the course of a year, regular climatic changes occur everywhere on Earth, to a greater or lesser extent. These changes are called seasons and they occur because Earth's rotation axis does not stand upright relative to our planet's orbit but tilts at an angle of 23.5 degrees.

THE POWER OF SOLAR ENERGY

Seasonal changes in climate result from the angle at which solar energy falls on the atmosphere, the land, and the oceans.

For example, let's select a region in the northern latitudes, roughly where Europe, North America, and Asia lie. Around June 21 in these latitudes, Earth reaches a stage in its orbit where the North Pole tips most fully toward the Sun. This point is known as the summer solstice, when "the Sun stands" before appearing to make its way south again.

This is the time when the Sun makes its longest arc across the sky. It rises in the northeast, stands high in the south at noon, and sets in the northwest. Its strength at noontime is palpable, pouring energy into the air, the ground, and bodies of water. The solstice marks the first day of summer,

although the warmest period usually comes a month or two later. This is because water in the ecosystem responds slowly to energy changes, taking on heat or giving it up only reluctantly. (Desert regions tend to respond more quickly to changes in the amount of sunlight.)

Six months later, around December 21, the situation is reversed. The North Pole now points away from the Sun. This is the time of the Northern Hemisphere winter solstice, when the Sun makes its shortest arc across

EARTH'S ORBIT

Earth's axis tips at an angle of 23.5° to the plane of its orbit. Thus during its annual waltz round the Sun, solar rays strike the ground at ever-changing angles. The varying amounts of solar energy produce the seasonal changes we experience—and also drive Earth's weather.

March 21:
northern spring
southern autumn

December 21:
northern winter
southern summer

Sun

June 21:
northern summer
southern winter

September 21:
northern autumn
southern spring

NIGHT AND DAY

Our daily experience is that of a fixed Earth at the center of the universe. So it's always a bit of a jolt to be reminded that we live on a ball of rock and water that spins once a day in the glare of a star.

ABOVE: Autumn's blaze of color marks an evolutionary adaptation. As sunlight lessens, broadleaf trees and shrubs stop producing chlorophyll, shut down photosynthesis, and drop their "solar panels" (leaves). By going into a form of hibernation, plants that originally evolved in a warm year-round climate are able to survive the cold of middle-latitude winter.

the sky, and its radiation strikes the northern latitudes obliquely, reducing its warming effects. Daylight may last only 8 hours instead of summer's 14 or 15 hours. Winter begins at this point although, as with summer, the coldest period comes several weeks later.

WHEN THE SUN IS OVERHEAD

Around March 21 and September 21, Earth is broadside to the Sun. On these dates—the equinoxes—the Sun is directly over the Equator, and day and night are of equal length. In the Northern Hemisphere, these dates mark the beginning of spring and autumn respectively.

Earth's distance from the Sun has little effect on the seasons. Earth is closest to the Sun in the first week of January and farthest away in the first week of July, but the difference

amounts to only a 7 percent change in the strength of the sunlight. This is swamped by the axial-tilt effects, which can vary by more than 200 percent from summer to winter.

THE SOUTHERN EXPERIENCE

In the Southern Hemisphere, the seasons are the opposite of those of the North. The climatic effects are the same, although the swings tend to be smaller because the greater amount of ocean in the southern half of Earth moderates the temperatures more.

The humid tropics, with their steady high temperatures, are as subject to seasonal changes as anywhere else. The regular alternation of dry and wet periods occurs in response to winds and ocean currents, which are influenced by seasonal changes that take place outside the tropics.

THE MOON

The Moon has accompanied Earth since the beginning. Earth had barely formed when a Mars-sized object struck it obliquely. The blow liquefied the impacting object plus most of Earth and sent a spray of vaporized rock into space. The spray cooled into a ring of rocky debris, which coalesced into the Moon in a few tens of thousands of years. Because it was small—2,160 miles (3,476 km) in diameter—the Moon cooled fairly quickly. This meant that compared with Earth (or even Mars) its geological "engine" ran down sooner and activity ceased not long after its birth.

THE MOTION OF THE MOON

The Moon circles Earth every 29.3 days, moving through a cycle of phases (see facing page). This cycle formed the basis of most early calendars and led to the 12-month year that is used throughout the world today.

When the Moon was young, it orbited much closer to Earth than its current distance of 238,856 miles (384,401 km). The gravity of the Moon raised tides on Earth, while Earth's gravity locked the Moon's rotation to keep the same side always turned toward us.

Lunar tides still raise the oceans twice a day. One bulge of water, on the side nearest the Moon, marks where its gravity is pulling the water away from Earth. The other bulge, on the opposite side of Earth, marks where the Moon is pulling Earth away from the water.

CRATERS AND SEAS

The lunar face is ancient, especially the lighter areas. These are the highlands, a craterscape of bewildering complexity. Violent impacts by asteroids and meteorites have rained upon the lunar surface for the past 4.6 billion years.

In the Moon's early days, impacts were larger and much more frequent. The biggest created wide basins, hundreds of miles across. Many of these have been filled by pools of dark

RIGHT: The face of the Moon has appeared essentially unchanged for more than half Earth's history. The Moon that shone on tyrannosaurs hunting at night would look completely familiar to today's backyard astronomers.

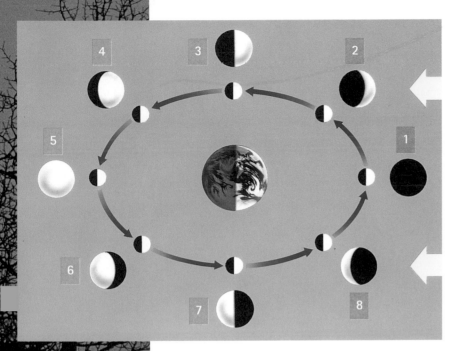

PHASES OF THE MOON

1 NEW MOON At New Moon the Moon lies directly sunward from Earth and we cannot see it because no sunlight falls on the side facing us.

2 CRESCENT Each evening after New Moon finds the Moon a little farther along its orbit, its crescent a bit wider, as it moves roughly its own diameter every hour.

3 FIRST QUARTER About a week after New, the Moon is one-fourth the way around, so the phase is called First Quarter. The Moon appears half-lit.

4 GIBBOUS After First Quarter the lit portion continues growing, and is described as gibbous.

5 FULL MOON At Full Moon the disk is completely lit. Standing opposite the Sun, the Full Moon rises as the Sun sets, and sets the next morning as the Sun rises.

6 GIBBOUS After Full Moon the lit portion starts to dwindle. The Moon rises later each night after sunset. It can often be glimpsed in the west soon after sunrise.

7 LAST QUARTER When the Moon has only one quarter of its journey around Earth to cover, the phase is called Last Quarter. The Moon now rises around midnight and again looks half-lit.

8 CRESCENT After Last Quarter the illuminated part shrinks to a crescent that becomes ever thinner as the Moon rises a shorter and shorter time before the Sun does. Finally, the Moon slips between Earth and the Sun and begins its age-old circuit again.

lava that oozed into them, starting about 3.9 billion years ago. The lavas flowed until about 2.5 billion years ago, when the Moon's volcanic activity ceased. Early observers mistook these lava-filled basins for dried-up ocean beds and called them "seas." They are more formally known as mare (pronounced MAR-ay, plural *maria*), after the Latin word for "sea." Today, the maria delineate what some people see as the face of "the Man in the Moon."

Impacts continued, however, and the more recent have left bright streaks called rays. These are splashes of rocky material that are flung out when meteorites make craters. Copernicus, an 800-million-year-old crater, has many rays, but the biggest and brightest rays of all belong to Tycho, which is a relative youngster at some 109 million years old.

The Moon now appears to be geologically dead, but six manned missions to the Moon (see pages 144–145) brought back hundreds of pounds of rocks that have done much to unfold lunar history and pin dates to events in its calendar. The immense age of the Moon's surface makes it a museum of solar-system history sitting on our doorstep. Study of the rocks brought back by Apollo continues today, as scientists investigate our part of the solar system, and how it fits into the overall picture.

ABOVE: A few days after New Moon, a bright crescent wraps around a ghostly blue disk: the "old Moon in the new Moon's arms." This "old Moon" is illuminated by earthshine—sunlight reflecting from the clouds and seas of Earth.

FEATURES OF THE MOON

The best time to observe the Moon by telescope is not at Full phase, because that is when the lighting is flat and unrevealing. You will see much more when sunlight is hitting the surface obliquely, throwing every feature into sharp relief.

THE YOUNG MOON

Roughly midway between New Moon and First Quarter, the line of lunar sunrise stands to the west of Mare Crisium. This lava-filled impact basin appears elongated north-south, but actually forms an oval extending east-west. The mare is some 300 miles (500 km) across, but the basin's outermost rim reaches twice that size. South of Crisium lies Mare Fecunditatis, another sheet of lava filling an impact basin. But this mare is much older and its outlines are not so distinct.

South of Fecunditatis, the rugged southern highlands appear. In this region the profusion of craters reaches saturation: any new impact would destroy an existing crater. This is the oldest terrain on the Moon and foreshadows what most of the lunar far side looks like.

FIRST QUARTER

At First Quarter, the eastern hemisphere of the Moon shows three additional "seas," including Mare Tranquillitatis, site of the first manned landing. (Unfortunately, the *Apollo* landers are much too small to be seen from Earth.) Mare Serenitatis displays an extensive ridge on its eastern expanse, buckled up as the floor of the basin subsided under the great weight of lava filling it. Crossing the mare is a bright ray, a streak of shattered rock thrown from the crater Tycho lying hundreds of miles to the southwest.

Mare Nectaris lies on the edge of the southern highlands. Unlike other seas, here the lava fills the impact basin only to its first "bathtub ring," a diameter of about 220 miles (350 km). Other rings, notably the Altai Scarp (Rupes Altai), stand in broken array around it, easy to see in the strong light of lunar morning. Nearby, craters such as Theophilus display complex interiors, with slumped rim walls, central peaks, and partly flooded floors.

WAXING BRIGHTER

After First Quarter, sunrise advances across the western hemisphere, home to some of the Moon's most interesting terrain, including giant Mare Imbrium, 750 miles (1,200 km) across. The impact that created the Imbrium basin affected nearly the entire Moon and its scars fall across much of the near side. The Imbrium impact thus marks a reference date in lunar history—3.84 billion years ago. The lava filling the basin is several hundred million years younger, however.

The Moon's greatest showpiece is the crater Copernicus, measuring 58 miles (93 km) rim to rim. Its insides are terraced with slumped portions of wall, its central peaks have dredged up material from deep in the Moon, and its floor was flooded in several stages. Watching this crater from the time its rim peaks first catch the rays of sunlight until they vanish 14 Earth days later is an unforgettable experience for any telescope owner.

FULL MOON

At Full Moon, the light from the Sun flattens all perspective, leaving only variations of light and dark caused by differences in age and composition. This is the time to trace the rays, splashes of debris that surround craters that are young, at least in lunar terms.

After Full Moon, features take on a different appearance with the light falling from a westerly direction. As sunset marches across the surface, shadows grow and hide the landscape for two weeks—until the Sun rises over them again, restarting the endless cycle.

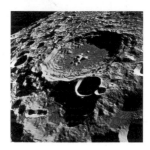

TOP: *This view from* Apollo 12 *shows the crater Reinhold in the foreground and Copernicus on the horizon.* ABOVE: *The terrain on the far side of the Moon largely resembles that around Daedalus, seen in this photo taken by* Apollo 11.

BELOW: *Twelve humans have strolled on the Moon, including Apollo 17 geologist–astronaut Harrison Schmitt, shown here in the Taurus-Littrow valley.*

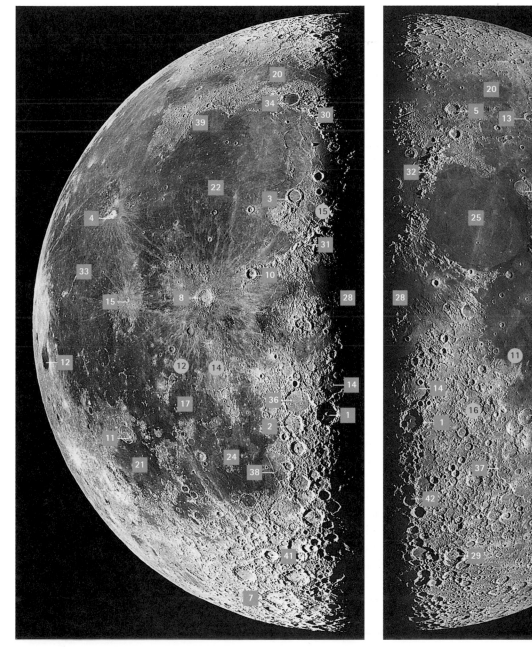

FEATURES OF THE MOON

The great variety of craters, mountains, and rays on the Moon could fill a lifetime of telescopic exploration. This view combines a photo taken at First Quarter (right) with one showing Last Quarter (above). Start observing with a low-power eyepiece and increase magnification slowly.

1. Albategnius
2. Alphonsus
3. Archimedes
4. Aristarchus
5. Aristoteles
6. Atlas
7. Clavius
8. Copernicus
9. Endymion
10. Eratosthenes
11. Gassendi

12. Grimaldi
13. Hercules
14. Hipparchus
15. Kepler
16. Langrenus
17. Mare Cognitum
18. Mare Crisium
19. Mare Fecunditatis
20. Mare Frigoris
21. Mare Humorum
22. Mare Imbrium

23. Mare Nectaris
24. Mare Nubium
25. Mare Serenitatis
26. Mare Smythii
27. Mare Tranquillitatis
28. Mare Vaporum
29. Maurolycus
30. Montes Alpes
31. Montes Apenninus
32. Montes Caucasus
33. Oceanus Procellarum

34. Plato
35. Posidonius
36. Ptolemaeus
37. Rupes Altai
38. Rupes Recta
39. Sinus Iridum
40. Theophilus
41. Tycho
42. Werner

● Apollo Landing Sites

SOLAR AND LUNAR ECLIPSES

A total eclipse of the Sun is one of the most eerie and awe-inspiring events you can see in the skies. To watch the Moon slowly turn broad daylight into night is a spectacle that no one who sees it ever forgets.

DAYTIME DARKNESS

The path of visibility for a solar eclipse may run for thousands of miles, but most people have to travel to see one because the track is often no more than 200 miles (300 km) wide. For those who make the journey, the sight is spellbinding. As the Moon makes it way across the Sun, it sends a chill down your spine.

During the partial phases (which last less than an hour), the darkening starts slowly, but the final stages pass swiftly. The temperature drops and daylight deepens to gloom. Once the total phase begins, the pale flower of the Sun's corona rings the black disk of the Moon. Stars and planets appear in the sky. Then, long before you would like it to end, the sky starts to brighten again. A portion of the Sun reappears and banishes the corona and the stars. The partial phases then unfold in reverse order as the eclipse comes to an end.

Solar eclipses occur because the Moon happens to be the right size to cover the Sun. At every New Moon, the Earth, Moon, and Sun line up approximately. But eclipses seldom take place, because the lunar orbit does not lie in the same plane as Earth's orbit around the Sun. Instead, it inclines by some 5 degrees. So at most New Moons, the Moon passes north or south of the Sun.

Depending on the circumstances, an eclipse of the Sun can be total, partial, or annular. In a total eclipse, the disk of the Moon covers the Sun completely; these are rare because the Moon is only just large enough to do the job. Partial eclipses—when the Moon covers part of the Sun—are considerably more common.

An annular eclipse is a particular kind of partial eclipse. This type of eclipse occurs because the Moon's orbit is elliptical, so its distance from Earth varies. If the Moon lies too far from Earth to cover the Sun completely, onlookers see a bright ring, or annulus, of uneclipsed Sun surrounding the Moon.

IN EARTH'S SHADOW

Eclipses of the Moon form the other half of the picture. They typically occur at the Full Moon preceding or following a solar eclipse, because that's when the Moon is most likely to pass through the shadow of Earth. This shadow extends behind Earth like a cone of darkness pointing at the stars.

A lunar eclipse is easy to see and lasts for several hours. If skies are clear, anybody on the night side of Earth can follow it. And it's safe to look at directly with the naked eye, binoculars, or a telescope. As the Moon edges its way into Earth's shadow, first of all you'll see it take on a dusky gray color, which then turns ruddy as the Moon glides into eclipse.

Some eclipses of the Moon are brighter or more colorful than others; dark ones often take place in years following large volcanic eruptions on Earth. The ruddy color occurs because our planet's shadow is faintly lit with sunlight that has filtered through the entire atmosphere—it is the light of all the world's sunrises and sunsets falling on the Moon at once.

Looking directly at the Sun can cause permanent damage to your eyes, so always use a proper solar filter when viewing the stages of a solar eclipse. You can hand-hold a filter that's designed to cover a telescope's aperture (see page 133).

SOLAR AND LUNAR ECLIPSES
From Earth, the Moon and Sun look about the same size. Thus the Moon can eclipse the Sun by passing before it at New Moon, and can itself be eclipsed at Full Moon if it enters Earth's shadow. The alignment of the three bodies is delicate, however, so most New and Full Moons pass without an eclipse.

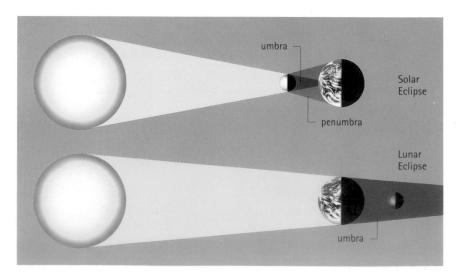

umbra

penumbra

umbra

Solar
Eclipse

Lunar
Eclipse

TOP LEFT: The color of a totally eclipsed Moon comes from sunlight refracting through Earth's atmosphere, staining the gray lunar dust with the hues of sunrise and sunset.

LEFT: A total eclipse offers the only chance to see the corona, the Sun's outer atmosphere, normally overpowered by the brilliant solar disk. This photograph captures the inner corona, a bright portion of uneclipsed Sun, and faint pinkish prominences—huge loops of glowing hydrogen— in the corona.

RIGHT: Time-lapse photos of eclipses are easy to take and very dramatic. You need a camera that can take multiple exposures, a wide-angle lens, and a tripod. The photographer of this image placed the partially eclipsed Moon at the upper left of the frame and snapped a short exposure. Then, without changing the camera's aim or advancing the film, he took an exposure every few minutes as the Moon slipped into total eclipse and emerged again before setting.

INSET: This photo shows a similar sequence of a partial eclipse of the Sun.

MARS

For all its small size, being roughly half as big as Earth (its diameter measures 4,217 miles, or 6,787 km), the planet Mars has certainly loomed large in our imaginations. This little red world draws our attention (and our spacecraft) as few others do.

Ancient civilizations associated Mars with a god of war or discord, perhaps as much in tribute to its varying brightness as its ruddy color. With the invention of the telescope, it was often seen as an inhabited "twin of Earth."

AN EXTREME CLIMATE

With clouds, storms, and seasons, Mars is the most Earth-like of the Sun's family. Its year lasts 687 Earth days, and by coincidence its day (known as the sol) is 24 hours 37 minutes long. Another near-parallel with Earth is the 25° tilt of Mars' axis. This gives the planet four distinct seasons, although its climatic extremes exceed anything we experience.

The Martian climate is best described as brutal. The atmosphere is 95 percent carbon dioxide. Carbon dioxide is a greenhouse gas, which locks heat in and helps keep a planet warm. But the Martian atmosphere is thin—less than 1 percent of Earth's—and offers only a small barrier to escaping heat. Martian surface temperatures barely reach 32°F (0°C) by day and drop to −190°F (−123°C) at night. In the arctic regions in winter, the atmosphere directly deposits dry ice on the polar caps.

Mars' elliptical orbit augments the seasonal changes. The planet is closest to the Sun in the southern summer, and sunlight then shines about 40 percent stronger than when Mars is at its farthest distance from the Sun.

DUSTY RED DESERT

The Martian world is a desert one, and the planet's ruddy color (detectable even by the naked eye) comes from its rusty, oxidized rocks and dust. A telescope shows an ocher-colored surface with darker markings. Once believed to be areas of vegetation, these are now known to be vast lava flows and boulder fields. Windblown sheets of fine dust and sand cause the changes in these markings that misled early observers into thinking they had seen evidence of plants or lichens.

Although thin, the Martian atmosphere experiences dust storms, some of which have hidden the planet for weeks on end. Ordinary

ABOVE: A dusty Martian sunset gives the sky a pinkish cast, just as it would on Earth, in this view taken by the Mars Pathfinder *lander. Clouds of water-ice crystals in the air make the Sun look hazy. When* Pathfinder *landed in 1997, it found a colder and somewhat less dusty Mars than that seen by the* Viking *spacecraft in the mid-1970s.*

RIGHT: About a hundred Viking *orbiter images went into this mosaic showing the region of Mars near Syrtis Major, the large dark area at the top right. This feature, the first to be identified in Earth-based views (by Christiaan Huygens in 1659), is a vast floodplain of basaltic lava.*

LIFE ON MARS?

We've long been susceptible to science-fiction stories about little green men from Mars, and a 1938 radio dramatization of H.G. Wells's novel *The War of the Worlds* had many Americans believing Martians had invaded the United States. The question of Martian life was reopened recently with the discovery of what appear to be fossil bacteria in a meteorite from Mars. Scientists think the rock is from Mars because gases inside it match analyses of the Martian air made by the *Viking* landers. (See also page 154.)

At present Mars probably can't support life. But for its first billion years or so it closely resembled Earth, with abundant water. Life may have begun on Mars—and then disappeared as the planet cooled and gave up most of its water to space. Whether life once existed there, or still survives, is a question that only extensive exploration is likely to settle.

RIGHT: *The northern polar cap of Mars (like the southern) can be seen in backyard telescopes (see page 134). Mostly water-ice mixed with dust, the cap sits on a "layer cake" of water-ice deposits that is several miles thick.*

clouds also form, often over the highest mountains a few hours after local sunrise. At night frost covers the ground in low-lying basins such as Argyre and Hellas, which are both ancient, weathered impact scars.

Water exists in the Red Planet's polar caps and in its atmosphere, and probably also lies in the ground as ice. Since the 1970s, scientists have seen small cyclical changes in the amount of water in the atmosphere; they think that Mars experienced vastly larger changes in the past.

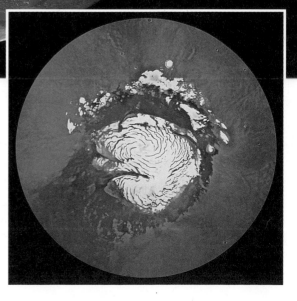

FEATURES OF MARS

Peering at Mars through a telescope imposes an extremely limited view of the planet. Even the sharp eye of the Hubble Space Telescope is unable to resolve any details on this little red world finer than several miles across. And the situation was a good deal worse decades ago, in the days before the advent of spacecraft and satellite telescopes.

SEARCHING FOR MARTIAN LIFE

At the beginning of the 20th century, Percival Lowell founded an observatory for the purpose of finding life on Mars, which he firmly believed existed. Lowell made countless maps and globes of Mars, showing it laced with thin canals, which he thought were enormous

LEFT AND BELOW: The Valles Marineris canyon stretches nearly 2,500 miles (4,000 km) and reaches depths of 4 miles (6 km). It runs radially away from the Tharsis volcanic region. Scientists think that activity in Tharsis broke open the crust along faults and landslides, then widened the canyon as ice washed out of the canyon walls.

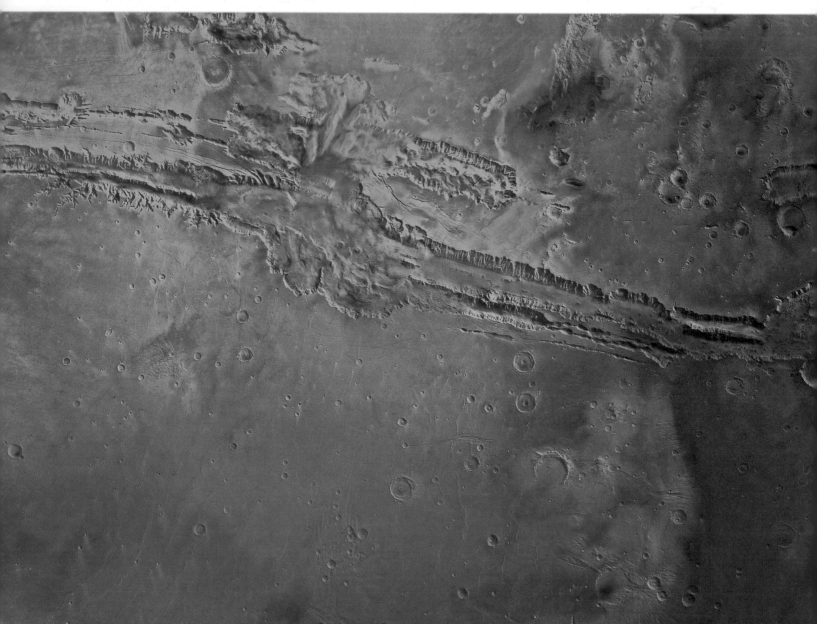

ABOVE: *Human beings excel at finding patterns in the natural world, but sometimes there's nothing really there. The long-famed "Martian canals" were optical illusions created by low-contrast details on the disk of Mars.*

RIGHT: *The Face is a mountain in Cydonia about a mile across. Much has been made of it by those who propose ancient Martian civilizations.*

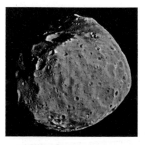

ABOVE: *A rock 7 miles (11 km) wide, Phobos orbits Mars along with its smaller sibling Deimos, 4 miles (6 km) across. These two Martian moons show many craters and are probably captured asteroids.*

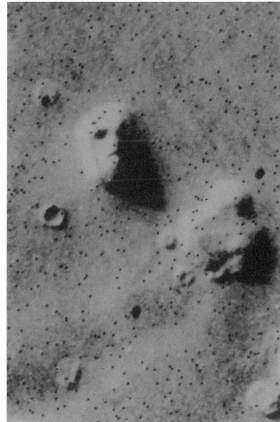

engineering projects designed to carry polar water to the barren regions around the equator. Lowell's canals were a great hit with the public but were scorned by professional astronomers as optical illusions. As it turned out, the astronomers were correct.

In 1965 the American *Mariner 4* spacecraft flew past Mars. It found a barren, almost lunar surface that was covered with craters. So a picture of the real Mars began to emerge, after centuries of wishful speculation. Follow-up probes—*Mariners 6, 7,* and *9, Vikings 1* and *2;* the Soviet *Phobos 1* and *2; Mars Pathfinder* and *Global Surveyor*—helped fill in the picture with details that would have astounded earlier generations of planetary astronomers. (See also pages 146–147.)

VOLCANOES AND WATER CHANNELS

In some respects, Martian landscapes parallel those on Earth. Mars has volcanoes: the towering Olympus Mons, 340 miles across and 15 miles high (550 km by 25 km) is just the largest of four giants in the Tharsis region. Mars has channels hundreds of miles long, down which enormous floods of water have raced, scarring the terrain and carving out streamlined islands. Great Martian canyons have opened. Valles Marineris yawns several miles deep and could span the United States from coast to coast. Landslides, fields of sand dunes, ice caps—all are familiar to us from terrestrial examples.

But Mars also differs substantially from Earth, especially in its craters. On Mars these are far more numerous—testimony to the fact that weathering takes place at a slower pace on the Red Planet. They range from simple bowls that are only a few hundred yards across to basins such as Hellas, which is 1,400 miles (2,200 km) in diameter and was formed by the impact of an asteroid.

And while much has been learned recently, Mars still has its mysteries. It displays two faces. One is its heavily cratered southern hemisphere, the surface of which is nearly as old as the oldest parts of the Moon. The other face is the vast northern lowlands, a younger and more lightly cratered area, with sedimentary deposits and extensive lava flows. What caused this giant dichotomy is unknown.

FURTHER MYSTERIES

Another mystery is where all the water on Mars has gone. The floodwaters that once eroded the channels are now nowhere to be seen. The polar caps are not thick enough to account for them, and while some water would have escaped into space, a much larger amount must have gone underground. Perhaps it underlies the polar regions, like the arctic permafrost on Earth.

And then there's The Face. In a region on Mars that is known as Cydonia, one particular peak was photographed by both *Viking* orbiters. Lighting conditions have lent the peak notoriety, making it roughly resemble a human face that's looking upward. The Face is surely a natural object, but perhaps it carries a message for us. Once it was canals, now it's faces—we should beware of optical illusions whenever we find patterns on Mars.

ASTEROIDS

In 1596 Johann Kepler announced, "Between Mars and Jupiter I shall place a planet." But for 200 years neither Kepler nor any other astronomer was able to discover an object to fill the gap that lies between Mars orbiting at 1.5 astronomical units and Jupiter at 5.2 astronomical units. It was not until January 1, 1801, that Giuseppe Piazzi discovered the first of the minor planets. He named the body Ceres.

Ceres was small, however, and its starlike appearance lent it the alternative name these bodies are still known by: "asteroid." Ceres is about 600 miles (1,000 km) in diameter, and it is the largest asteroid.

MINOR PLANETS

More discoveries followed that of Ceres, although the work proceeded slowly when astronomers had to meticulously compare the view in a telescope with star maps compiled by hand. Roughly five minor planets were found every year until the end of the 19th century, when photography boosted the rate significantly. Today astronomers use both film and electronic detectors to find new objects and keep watch on known ones. At present more than 35,000 minor planets are known.

Most minor planets orbit within the main belt, the region between Mars and Jupiter, and their orbital periods last several years. Those closer to the Sun have compositions that are more purely rocky, while those nearer Jupiter

contain additional organic and carbon-related compounds. A few asteroids have surfaces as bright as weathered concrete, but most are dark, rather like asphalt pavement.

The main belt is not a smooth band, however. Gaps exist where asteroids would orbit in periods that are simple fractions of Jupiter's 12-year period. These gaps help explain why the minor planets have remained minor. When the solar system took shape, Jupiter's enormous gravity prevented any planet from forming in this region, and today's asteroid belt is the winnowed remnant of a far richer population that once orbited there.

BELOW: Collisions in the asteroid belt create dust that drifts in toward the Sun. Some dust drifts all the way around Earth's orbit, as seen in this computer-simulated view.

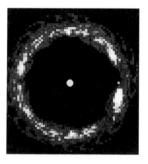

ASTEROID ORBITS

The elliptical orbits marked on the diagram belong to some of the near-Earth asteroids— ones that cross the orbit of Earth. The vast majority of asteroids, however, orbit in the main belt. The Trojans are two clusters of asteroids that lie in the orbit of Jupiter.

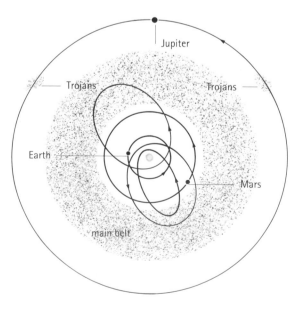

FAR LEFT: In October 1991, on its cruise toward Jupiter, the Galileo *spacecraft passed asteroid 951 Gaspra, which measures 11 by 7 by 6 miles (18 by 11 by 9 km) and has a "day" lasting some 7 hours.*

LEFT: Galileo *flew past asteroid 243 Ida in August 1993 and found a 33 mile (53 km) long piece of rock with an old, heavily cratered surface. Ida's moonlet, Dactyl, which is 1 mile (1.5 km) in diameter, came as a surprise.*

ABOVE RIGHT: Asteroid 253 Mathilde (on the left) is seen here to scale with Gaspra (center) and Ida (right). Mathilde is scarred by more large impacts than the other two, perhaps reflecting a more violent history. The Near Earth Asteroid Rendezvous spacecraft, which imaged it in June 1997, also found Mathilde was about as dark as a piece of tar.

Not all asteroids orbit within the main belt. Collisions among them, and again perturbing tugs from Jupiter, have dispatched asteroids and fragments on diverse courses that cruise the solar system. Jupiter has two clutches of asteroids traveling with it in its orbit, one 60° ahead and one 60° behind. They are known as the Trojan asteroids. Other asteroids have trajectories that approach Earth or even cross its orbit, with some risk of collision. (Most scientists believe that the dinosaurs were wiped out in the aftermath of a small asteroid striking Earth 65 million years ago, and that nothing but luck stands in the way of another impact occurring at any time.)

It is probably impossible to know exactly how many asteroids exist. Astronomers think they have found all the main-belt asteroids that are 60 miles (100 km) or more in diameter, and perhaps half of the 6 mile (10 km) objects. For smaller ones, the percentage known is still only a token. Estimates suggest that there are about a million asteroids that are ½ mile (1 km) in diameter, but collisions are always changing the population.

PROBING THE PAST

Planetary astronomers are interested in asteroids because these objects provide a view into the distant past. They preserve relatively intact one part of the solar nebula from which the planets formed. Some asteroids are primitive shards of rock, little altered from the moment they cooled 4.6 billion years ago. Others are fragments of larger bodies that may have undergone some geological evolution.

HOW ASTEROIDS ARE NAMED

When an astronomer discovers an asteroid, he or she earns the right to suggest a name. Early discoveries bear names from classical mythology, but antiquity has been ransacked by now and thousands more objects still await names. So today we have asteroids with names such as 6000 United Nations, 2985 Shakespeare, and even 4659 Roddenberry (of *Star Trek* fame). And of course there's 4147 Lennon, 4148 McCartney, 4149 Harrison, and 4150 Starr.

METEORS AND METEORITES

On any clear dark night, even in the city, you probably can spot a "shooting star" if you watch the skies for a while. From a rural location, you will see a dozen an hour, even more at certain times of the year.

A shooting star, or meteor, is caused by a small piece of interplanetary material falling into Earth's atmosphere at high speed. The piece of debris, called a meteoroid, enters the atmosphere at a speed ranging from about 6 to 45 miles (10 to 70 km) per second. Friction with air molecules heats and vaporizes it, resulting in the bright streak we call a meteor. Some meteors leave faintly glowing trains— trails of dust or ionized gas—behind them.

GLOWING DEBRIS

Even the brightest meteors, called fireballs (or bolides if they make a final burst), are remarkably small and fragile. Most are as tough as cigarette ash, and one the size of a grain of sand would be bright enough to elicit an "Ooooh!" from onlookers. According to planetary scientists, most meteors are debris shed by comets (see pages 98–99), which are a mixture of ices and dust. When a comet nears the Sun, its ices evaporate, freeing the dust, which travels along the comet's orbit for a while.

At certain times of year, Earth sweeps through the trail of dust from a comet, and a meteor shower occurs. The Perseids of August are a well-known shower. Meteor showers take their name from the constellation the meteors appear to come from, and several dozen showers recur every year (see page 121).

Meteors that do not appear to belong to any shower are called sporadics. Most of them probably belonged to one shower or another long ago but the showers are now so depleted as to be unrecognizable.

Earth receives 100 to 1,000 tons of meteoritic debris every day. This sounds like a lot, but it is spread over the entire Earth. Most meteors burn out and their minute residue slowly drifts down through the atmosphere. But not all interplanetary material arrives so gently.

When asteroids collide in space, fragments fly, and some make their way to Earth. Any fragment retrieved intact is called a meteorite.

There are dozens of different kinds, weighing from a few dozen ounces to many tons.

Really big meteorites don't survive the landing, however. Northern Arizona's Meteor Crater is about 3,600 feet across and 600 feet deep (1,100 by 180 m). It was formed when an iron meteorite about 150 feet (45 m) in diameter struck with a force of 15 megatons of TNT. Only small fragments have been found.

ABOVE: *This fireball's path took it across the spiral galaxy NGC 253 in the constellation of Sculptor. Then, as it brightened to rival the Full Moon, it raced out of the camera's field of view.*

ROCK SAMPLES FROM SPACE

Meteorites are a wonderful gift to the planetary scientist, since each is a free sample of rock from space, delivered without the complication and expense of a mission. A dozen pieces of Mars are now sitting in various terrestrial laboratories, including the famous ALH84001 meteorite, which may contain fossil traces of ancient Martian life (see pages 76 and 154).

Most meteorites, however, are samples of the asteroid belt and they reflect the variety that exists there. Stony meteorites account for 95 percent of those that have been found, nickel-irons make up a bare 4 percent, and stony-irons the remainder. Scientists study these rock samples in order to learn about conditions in the early solar nebula and how it changed to become the planetary system of today.

JUPITER

The ancients chose Jupiter's name more wisely than they knew. This bright planet, which takes 12 years to circle the sky, has been associated in many mythologies with the most powerful of the deities. Perhaps it was the symbolism of spending a year in each constellation of the zodiac that was particularly impressive.

MONARCH OF THE SOLAR SYSTEM

In any case, Jupiter also ranks as the king of the planets on other grounds. It has a diameter of 89,400 miles (143,800 km) and weighs as much as 318 Earths. In fact, it weighs more than all the other planets put together. Its immense gravity directs the fate of many comets, and it can send asteroids careening through the solar system. It governs a family of 16 moons, three of which are larger than our own Moon.

Jupiter is also a gas-giant planet, with a composition and structure that is radically different from the four terrestrial planets. Its makeup is remarkably like that of the Sun—hydrogen accounts for 80 percent, helium 19 percent, and the rest consists mainly of water vapor, methane, and ammonia. Jupiter's rapid spin—less than 10 hours—makes it flatten at the poles and bulge at the equator, giving the planet a distinctly oval appearance.

The "surface" of Jupiter consists of layers of cloud standing near the top of an immense atmosphere thousands of miles deep. The colors and banding seen in photographs result from chemical reactions in the ammonia and methane. Below the top of the cloud deck the atmosphere remains gaseous to a depth of perhaps a few hundred miles.

Scientists think that beneath this gaseous layer increasing pressure and temperature turn the hydrogen into a molecular liquid that would behave rather like a hot mush. Such a layer could extend as much as 13,000 miles (20,000 km) into the planet. At that depth, the atmosphere undergoes another change, transforming into a metallic hydrogen that is highly conductive electrically. Finally, Jupiter's core is probably somewhat similar to a big terrestrial planet—a molten ball of silicate rock that is several times the mass of Earth.

Jupiter gives off more energy than it receives from the Sun because it is still cooling off from its formation. Electrical currents generated in the molten core and surroundings produce a large magnetic field that reaches far out into space. The field deflects and channels electrons and protons from the Sun into a complex shell and bombards the inner moons with particles, while trapping radiation near the planet. (The radiation belts, similar to those on Earth but vastly stronger, would be lethal to humans and can damage spacecraft electronics.)

Telescopes and spacecraft show that Jupiter's face is strongly banded, reflecting winds that blow east and west at speeds of several hundred miles per hour. How deep these winds go has not been established, but working from the *Galileo* probe's results of December 1995, scientists think that the winds probably extend to depths of thousands of miles.

THE GREAT RED SPOT

Besides the bands, many semi-permanent features in the atmosphere have been identified by planetary scientists and are tracked by both professionals and amateurs. Most famous is the Great Red Spot. This large storm in Jupiter's mid-southern latitudes is twice the size of Earth and has been observed for several hundred years. Scientists have seen it absorb small disturbances that crashed into it, yet how the Great Red Spot formed is still a mystery.

RIGHT: The gases in Jupiter's atmosphere change color in chemical reactions driven by sunlight. The Jovian day spins the dark belts and light zones into streaks and disturbances, driven by winds that blow in alternating directions parallel with the equator.
INSET: A false-color image of Jupiter's tenuous ring system. The particles are mostly rock dust, whereas those of Saturn's rings are largely water-ice.

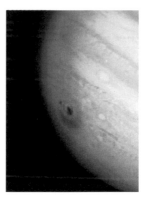

ABOVE: Comet Shoemaker–Levy hit Jupiter in July 1994 and this Earth-sized scar was caused by one of some 22 fragments. If the comet had slammed into Earth it would have killed most living things.

THE LARGEST PLANET

Jupiter consists almost entirely of hydrogen and helium, and at its center lies a core of molten rock, perhaps several times Earth's mass. Sometimes called "a failed star," Jupiter would need about 100 times more mass to become even the smallest type of star.

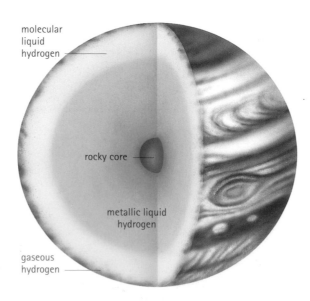

molecular liquid hydrogen

rocky core

metallic liquid hydrogen

gaseous hydrogen

THE MOONS OF JUPITER

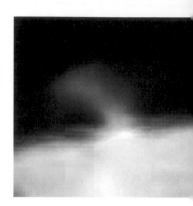

ABOVE: *In 1986 the* Galileo *spacecraft photographed a volcanic geyser on Io as it shot a blue jet of sulfur dioxide gas about 60 miles (100 km) high. Eruptions like this, which occur constantly on Io, have ensured that the moon's surface is the youngest of any in the solar system.*

LEFT: *Io is often compared to a pizza. In this contrast-enhanced photo taken by the* Galileo *orbiter, the black and red areas are the most recent volcanic deposits—probably no more than a few years old.*

When Galileo Galilei made the first telescopic observations of Jupiter in January 1610, he spotted four pinpoints of light accompanying the planet. Following their movements for weeks, he realized that they were orbiting Jupiter just as the Moon orbits Earth. A devoted Copernican at a time when this involved personal risk, Galileo was delighted to discover a solar system in miniature—and one that was operating on the Copernican model.

Galileo's discovery can be duplicated by anyone today with a pair of binoculars (see page 124), but our knowledge of the Jovian family of moons is much greater than his. We know of 16 moons now, and have close-up spacecraft portraits of the majority of them. The most recent images were taken by the *Galileo* orbiter, which has been prowling the Jovian realm since early 1996 (see page 147).

GALILEO'S MOONS

The four moons that were discovered by Galileo Galilei are the largest. From Jupiter outward, they are Io (EYE-oh), Europa, Ganymede, and Callisto. Our Moon is slightly larger than Europa (1,950 miles; 3,138 km), but smaller than Io (2,256 miles; 3,630 km), Callisto (2,983 miles; 4,800 km), and Ganymede (3,270 miles; 5,262 km). Together, these are known as the Galilean moons, and each has its own geological character.

RIGHT: The smallest of the Galilean moons, Europa may be the most intriguing. Its surface is nearly pure water-ice, a shell 10 to 60 miles (15 to 100 km) thick. The shell has cracked in many places because while Europa's rocky core rotates synchronously, keeping one face turned toward Jupiter, the ice shell does not.

BELOW: The farthest of Jupiter's moons, Callisto (left) is a mixture of ice and silicate rock. Next inward, Ganymede (lower right) is the largest of the moons and is also about half ice and half rock. Europa (center) and Io (far right) have greater proportions of rock—in fact, Io has no ice at all.

Io is the most volcanically active body in the solar system. As it circles Jupiter, it is tugged by the gravity of Jupiter, Europa, and Ganymede. This tidal flexing keeps Io's interior molten, with repeated eruptions in several locations (many of them named for fire-gods in mythology). Io is so active that its blotched, sulfur-coated surface shows no impact craters, and scientists think its upper layers have turned themselves inside out (via eruptions) every few million years.

Europa poses many enigmas. It has a smooth, bright crust that is scarred with a network of darker lines, like marks drawn on a billiard ball. Craters are few, implying the surface is young. From Europa's density, scientists believe that it has a rocky core and a deep ocean of water or slush, covered with the icy skin seen by spacecraft. The dark lines resemble freshly frozen openings in the polar ice pack on Earth. If Europa does have a global ocean, it might harbor life, although this is no more than wild speculation at present (see page 155).

Ganymede contains a mixture of rock and ice. Its face, rich in water-ice, has many craters. Yet it also has regions of grooved terrain, lighter in color and younger. Scientists are not sure what caused the grooves, but an expansion of Ganymede's interior as it melted throughout could have generated these volcanic/tectonic features. The activity apparently continued for some time before it froze in place.

Callisto is the least altered of the Galilean moons, its densely cratered surface resembling a firing range. The moon appears generally dark with silicate "dirt," but most of the craters have bright interiors, ray patterns, and exposed fresh ice. At least one large impact basin—Valhalla, 900 miles (1,500 km) across—shows where an asteroid or large comet struck the moon.

LESSER SATELLITES

The remaining worlds within the Jovian system tend to be heavily cratered. The ones that are closest to Jupiter are the most rocky, while those on the fringes have more ice in their make-up. The outermost four moons revolve around Jupiter in a backward direction and are probably captured asteroids.

SATURN

Saturn was the last of the planets known to antiquity. As it crept around the sky, taking 29½ years to complete a circuit, it seemed to embody the infirmities of old age. So the planet was labeled Saturn—the father of the gods in Greek and Roman mythology.

After the invention of the telescope in the 1600s, Saturn became the showpiece of the solar system. While the other three gas-giant planets also have their rings, Saturn's are the only ones that can be seen easily from Earth.

A LOW-DENSITY GAS GIANT

In many ways, Saturn could be considered a scale model of Jupiter. It has a smaller diameter (75,000 miles, or 120,660 km) and it is considerably lighter, with a mass equal to 95 Earths. All the gas giants have low densities compared with the terrestrial planets, and Saturn's is the lowest of all. The planet's density is actually less than that of water, and therefore it would float, if one were able to find a tub large enough to accommodate it.

Saturn's composition roughly parallels that of the Sun and Jupiter (see pages 58–59 and 84–85): 88 percent hydrogen, 11 percent helium, and small amounts of methane, ethane, and ammonia. Chemical reactions by the latter three cause Saturn's tan color and faint banding. A spin of 10 hours 12 minutes has given Saturn an even wider girth than Jupiter, and its winds scream 900 miles (1,500 km) per hour near the equator, four times the speed of Jupiter's winds.

Scientists think that Saturn's inner structure also resembles that of Jupiter. A layer of clouds covers a thick layer of fluid hydrogen that grows hotter and denser the farther it is from the surface. This probably becomes metallic about 20,000 miles (30,000 km) down. The core is thought to be a molten silicate ball weighing a dozen or more Earth masses.

PICTURING THE RINGS

The rings of Saturn span 170,000 miles (270,000 km) and tilt 29° to Saturn's orbit. They are no more than a few hundred yards

BELOW: This false-color image spans Saturn's rings from the inner C ring (bluish color at right), through the B ring (warm tints merging into greens and blues), out to the Cassini Division (dark with bright streaks). Outside the Cassini Division lies the A ring. While ring particles are mostly ice, they contain rocky impurities that alter their colors and which may come from micro-meteorite dust embedded in them or coating their surfaces.

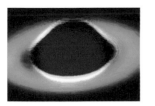

ABOVE: An infrared image made from Earth highlights the compositional difference between Saturn's disk and its rings, the methane-rich atmosphere showing red, and the icy rings blue.

RIGHT: In this image, snapped by Voyager 2, scientists have "turned up the contrast" so that Saturn's banded clouds can be easily seen. Also shown are the moons Tethys, Dione, and Rhea, and Mimas as a dark spot on Saturn's disk.

RIGHT: Jupiter's atmosphere is always stormy, while Saturn's seldom displays much activity. At roughly 30-year intervals, however—most recently in 1990—a storm breaks out near Saturn's equator. This false-color photo taken by the Hubble Space Telescope shows a large white cloud of ammonia-ice crystals. Starting as a spot, the cloud grew until it was easily visible in backyard telescopes from Earth, but after several months it disappeared.

thick, but can be seen in backyard telescopes (see page 134). The rings vanish from view for several hours roughly every 15 years, when Earth passes through the plane of the rings.

Before the *Voyager* spacecraft flybys that took place in 1980 and 1981, scientists thought they had a good picture of the ring system. They knew of three main rings. The outer was called the A ring. The dark, empty Cassini Division, 2,600 miles (4,200 km) wide, separated this from the wider and brighter B ring. And the thin and dim C (or crepe) ring lay within that.

RINGLETS WITHIN RINGLETS

Voyager's amazingly detailed photos therefore came as quite a shock. They showed more rings outside the main ones, and revealed that the three "classical" rings were made of ringlets upon ringlets, almost too many to count. Even the Cassini Division had loads of material in it. The new picture revealed a vast array of icy particles moving in enormously complicated ways—a puzzle scientists are still unraveling.

The ring particles are "hailstones" the size of golf balls, trucks, and houses. The rings are probably no more than 10 to 100 million years old, which is short-lived in solar-system terms. They exist because one or more moons collided or came close enough to Saturn for its gravity to break them apart. Scientists think that in a few tens of millions of years the rings will disappear as collisions between particles slow their orbits and the remains fall into Saturn.

THE MOONS OF SATURN

Like Jupiter, Saturn has collected a diverse family of moons, 18 of them in all. They consist primarily of rock and water-ice—and at Saturn's distance from the Sun, ice behaves as rigidly as steel—with craters dominating the surfaces of most of them. All but the outer two moons orbit fairly close to Saturn's equatorial plane. They range in size from tiny Pan, at 12 miles (20 km) across, up to Titan, which is 3,200 miles (5,150 km) in diameter.

SMOGGY TITAN

Titan is larger than Mercury and has an atmosphere that is 50 percent heavier than Earth's. When the *Voyager* spacecraft flew by, they saw nothing of Titan's surface. While the atmosphere is primarily transparent nitrogen, it also contains methane, ethane, acetylene, ethylene, and a variety of other compounds. Exposed to sunlight, these produce a photochemical smog resembling the smog that forms over Los Angeles on a bad day. The result is that Titan's surface features are largely hidden, which is a shame: scientists think that at Titan's temperatures (−292°F, or −180°C), pools of liquid methane may exist, along with methane clouds, rain, and snow, because methane takes on the various forms that water does on Earth.

Planetary scientists are eagerly awaiting the arrival of the *Cassini* spacecraft at Saturn in 2004, when it will drop an instrument probe (named *Huygens*) on Titan. Falling through the atmosphere, *Huygens* will analyze the gases and clouds, and measure wind speeds. Assuming that the probe survives, it will then send data from the moon's surface. (See also page 153.)

OTHER FAMILY MEMBERS

Saturn's oddest moon may well be Iapetus (eye-APP-uh-tuss), which is 907 miles (1,460 km) across. Like our Moon and most other satellites, Iapetus keeps one side turned toward the planet it orbits. What makes this moon unusual is that the forward-facing hemisphere is jet-black in color, while the one that is trailing is snow-white.

The white material appears to be ice, but the nature of the jet-black material remains a mystery. It may be dust that has been chipped off Phoebe (the next moon outward) and swept up by Iapetus—or it could have erupted through some internal process.

Hyperion is too small a moon for its gravity to overcome its irregular shape: it measures 255 by 162 by 137 miles (410 by 260 by 220 km). Heavily cratered, Hyperion does not orbit with its longest axis aimed toward Saturn, which would be its natural state. Planetary scientists think that an asteroid or comet may have hit the moon recently enough that it has not yet settled back into alignment.

Enceladus (en-SELL-ah-duss) appears to be the most geologically active of Saturn's moons, with the activity being driven by heat produced inside Enceladus by repeated gravitational tugs from its neighbor moon Dione. Enceladus has a diameter of 300 miles (500 km) and a density indicating that it consists largely of water-ice. It displays a bright surface with a mixture of old, well-cratered areas and newer terrain that is grooved and fissured.

The moon Mimas (MYE-mas) is 242 miles (390 km) in diameter. The brightness of its surface and its low density indicate an icy composition. Unlike Enceladus, however, it appears to be old and heavily cratered, which is a common fate in the solar system. Among its craters is a particularly large one, which gives it an uncanny resemblance to the Death Star in the movie *Star Wars*!

RIGHT: A composite of Saturn's larger moons, thanks to Voyager 1. Dione is at the front with Tethys to its right and Mimas at far right. To the left of Saturn's rings lies Enceladus, then Rhea. Enigmatic Titan, in its smog, stands at the upper right.

ABOVE: Like most of the Saturnian moons, Mimas is an iceball with a heavily cratered surface that has passively recorded impacts for billions of years.

LEFT: Geological activity on Enceladus (shown here in false color) has formed smooth plains and grooves that have erased some of its old cratered terrain. While the surface is highly reflective (indicating almost pure ice), scientists think the "lava" is a mix of ammonia and water, and estimate that the crater-free regions are probably no more than 100 million years old.

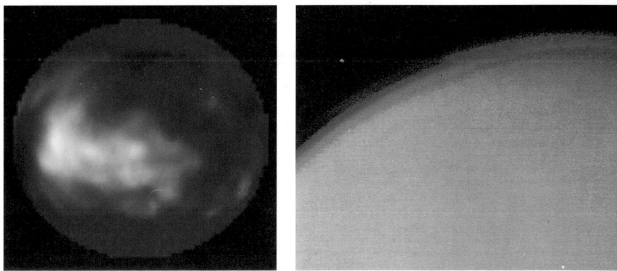

RIGHT AND FAR RIGHT: Titan is blanketed by smog (far right) but by using an infrared wavelength where the smog partly cleared, scientists with the Hubble Space Telescope were able to dimly image the surface (right). They found an Australia-sized "continent" and signs of possible impact basins. With all its organic compounds, Titan should prove a great source of pre-biological chemistry when the Cassini and Huygens spacecraft arrive in 2004.

Uranus

William Herschel stumbled upon Uranus one evening while he was surveying stars in the constellation Gemini. The date was March 13, 1781, and Uranus thus became the first planet to be found in modern times—and the first addition to the solar system since antiquity.

A Blue-green, Tilted World

Like its neighbors, Uranus is a gas-giant world. Smaller than Jupiter or Saturn, it has a diameter of 31,765 miles (51,120 km), nearly four times the size of Earth. It weighs as much as 14 Earths, and mimics the Sun in composition: hydrogen and helium comprise almost all of Uranus. Traces of methane give Uranus its soft blue-green color, vivid even in a backyard telescope.

Astronomers began suspecting something unusual about Uranus when they noted the orbits of its moons, which were inclined almost at right angles to Uranus' orbit. Today planetary scientists have indeed established that the tilt of Uranus' axis is a startling 98 degrees. This is a world orbiting on its side.

Voyager 2 Findings

The view from Earth doesn't reveal much of Uranus, so little was known about it until 1986, when the *Voyager 2* spacecraft sped past at close range. At the time Uranus was pointing its southern pole toward the Sun. As luck would have it, the planet presented a featureless appearance. This greatly frustrated scientists, who had been hoping to study features similar to those on Jupiter or Saturn. They did, however, determine that the Uranian day lasts 17 hours 15 minutes.

The spacecraft also provided information about Uranus' internal structure. The outer layer reaches inward about one-fourth of the radius, mixing hydrogen, helium, and methane. Below lies a hot slush of water, methane, and ammonia, combined with rocky components. The planet's magnetic field tilts relative to the planet's axis by some 59°—and is offset so it does not pass through the planet's center. Scientists guess that a big impact early in Uranus' history may have skewed the field.

Narrow and thin, the Uranian rings are kept in place by the gravitational effects of tiny moonlets. A typical ring particle is half a yard or smaller in size, and lasts perhaps 500 years before being destroyed when it collides with other particles or spirals into Uranus. Scientists think the particles were created by collisions between larger pieces within the rings and by impacts on nearby moons.

Five moons were known before *Voyager*, and the spacecraft added ten. (Recently, two more have been found from Earth.) Miranda is the most remarkable of all. The smallest of the original five, it is only 292 miles (470 km) in diameter. Besides the expected craters, Miranda showed many puzzling surface features. Among these are a 6 mile (10 km) cliff at the edge of a fault zone, and at least three huge areas with roughly concentric grooves. One possibility is that Miranda has been smashed apart and reassembled, and the grooves mark where heavier pieces have sunk.

No spacecraft will be visiting Uranus in the near future. Reviewing old observations made from Earth, however, scientists note that the planet tends to display banding at equinox, when the Sun is over its equator. *Voyager* visited Uranus one year past its solstice, when the Sun was still over its pole. Its next equinox will take place in 2007, so between now and then, some interesting features may emerge.

LEFT: *When Voyager 2 flew by in 1986, Uranus' bland appearance frustrated scientists hoping to map winds and track storms. Clouds do exist but are hidden below the bluish haze, which is caused by traces of methane in the hydrogen atmosphere.*

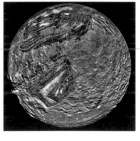

ABOVE: *Consisting largely of water-ice and coated with dust, the moon Miranda may have formed from fragments of one or more other moons.*

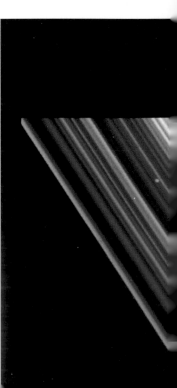

RIGHT: Cold, high-altitude hazes made of methane turn the edges of Uranus' crescent disk white. The Voyager 2 craft snapped this photo as it left the Uranian system and headed for Neptune, the next planet on its journey.

BELOW: The epsilon ring, the outermost of Uranus' ten rings, is not perfectly circular, which may indicate it formed in the geologically recent past. This computer image shows its structure on opposite sides of the planet but at the same distance from its center. The redder areas are where there is less ring material.

INSET: The rings' coal-black surface absorbs heat from the Sun, making them more visible than Uranus itself in some infrared wavelengths.

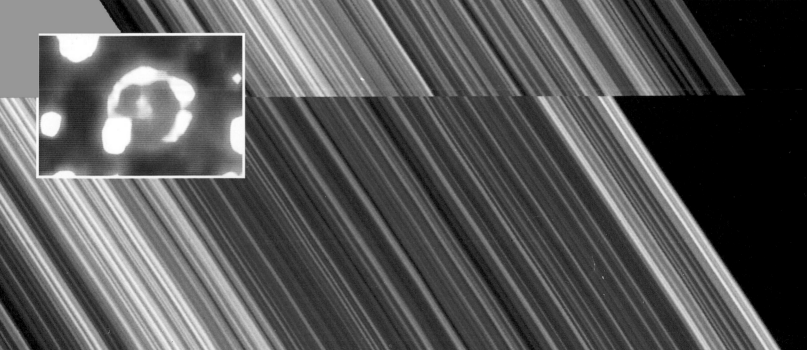

NEPTUNE

The ancient god of the sea and a blue-green planet rich in water make a fine association. However, in 1846 when the planet Neptune was named, its real nature was unknown.

THE SMALLEST GAS GIANT

Neptune has a diameter of 30,760 miles (49,500 km) and a mass equal to about 17 Earths. Although the planet is still making its first "known" orbit and won't reach the position where it was discovered until 2011, astronomers have calculated that it takes 164.8 years to complete one circuit and that its axis inclines 28.8° to its orbit. Of Neptune's retinue of eight moons, only two were known before *Voyager 2* paid a visit in 1989. *Voyager* also delineated the planet's five dark rings.

Hydrogen and helium dominate Neptune's makeup, along with methane, water, and other compounds. Methane lends it the blue color seen in a telescope and in spacecraft images. Scientists believe Neptune's internal structure broadly resembles that of Uranus, with layers of hydrogen and helium becoming steadily hotter toward the center. Neptune's greater density hints that the silicate-iron core at the center is somewhat bigger than the core of Uranus. (Neptune also has a weak magnetic field, tipped about 55° to its axis, but it is not displaced like the magnetic field of Uranus.)

At Neptune *Voyager* saw atmospheric bands, bright ammonia-ice clouds, the Great Dark Spot—a storm the size of Earth—and the Scooter—a fast-moving white cloud. Winds in the equatorial zone fly westward at 900 miles (1,500 km) per hour, while the planet spins on its axis every 16 hours 6 minutes. Neptune radiates more than twice as much heat as sunlight provides and this drives much of the activity.

Since the *Voyager* visit, the Hubble Space Telescope has recorded the disappearance of some features, notably the Great Dark Spot, and the emergence of new features.

INTRIGUING TRITON

Neptune has one large moon, Triton, which is 1,680 miles (2,706 km) in diameter. Triton revolves around Neptune in a backward direction, the only large moon in the solar system to do so. It is a mixture of ice and water, covered with pinkish frosts of nitrogen and methane condensed from the atmosphere. It has geysers that shoot dark smoke several miles into the sky and leave murky streaks on the surface. Triton has many impact craters, as well as a region nicknamed the "cantaloupe" terrain that may have resulted from gigantic frost heaves.

TRITON'S PROBABLE ORIGIN

Triton's odd orbit, and the elongated orbit of Nereid, the second largest moon, suggest that, rather than forming with Neptune, both were captured by its gravity. If so, they probably originated in the Kuiper Belt, a region that starts roughly at Neptune and extends outward a thousand astronomical units or more.

Planetary scientists also suspect that what they have found out about Triton foreshadows what they'll discover at Pluto (see pages 96–97), as both may have Kuiper Belt origins.

ABOVE: Neptune's largest moon, Triton is a complex world. Traces of methane color the pinkish frost cap of nitrogen-ice, which migrates around the moon by means of the atmosphere.

FINDING NEPTUNE

Unlike Uranus, which William Herschel came across by accident, the discovery of Neptune followed calculated predictions. Irregularities in the motions of Uranus led Urbain Le Verrier (above left) in France, and John Couch Adams (above right) in England, to determine the probable whereabouts of an unknown planet. Neither knew of the other's work, but their solutions more or less agreed.

British astronomers were slow in following up Adams's work, but Urbain Le Verrier sent his prediction to Johann Galle and Heinrich d'Arrest at the Berlin Observatory, who found Neptune within half an hour on September 23, 1846.

RIGHT: A giant storm on Neptune, the Great Dark Spot disappeared between the 1989 Voyager encounter (which produced this image) and 1994, when Hubble Space Telescope images showed it was gone. INSET: Voyager 2 also saw several bright cloud features and a smaller dark spot in the southern hemisphere that was dubbed D2, seen here near the lower-right edge of the planet.

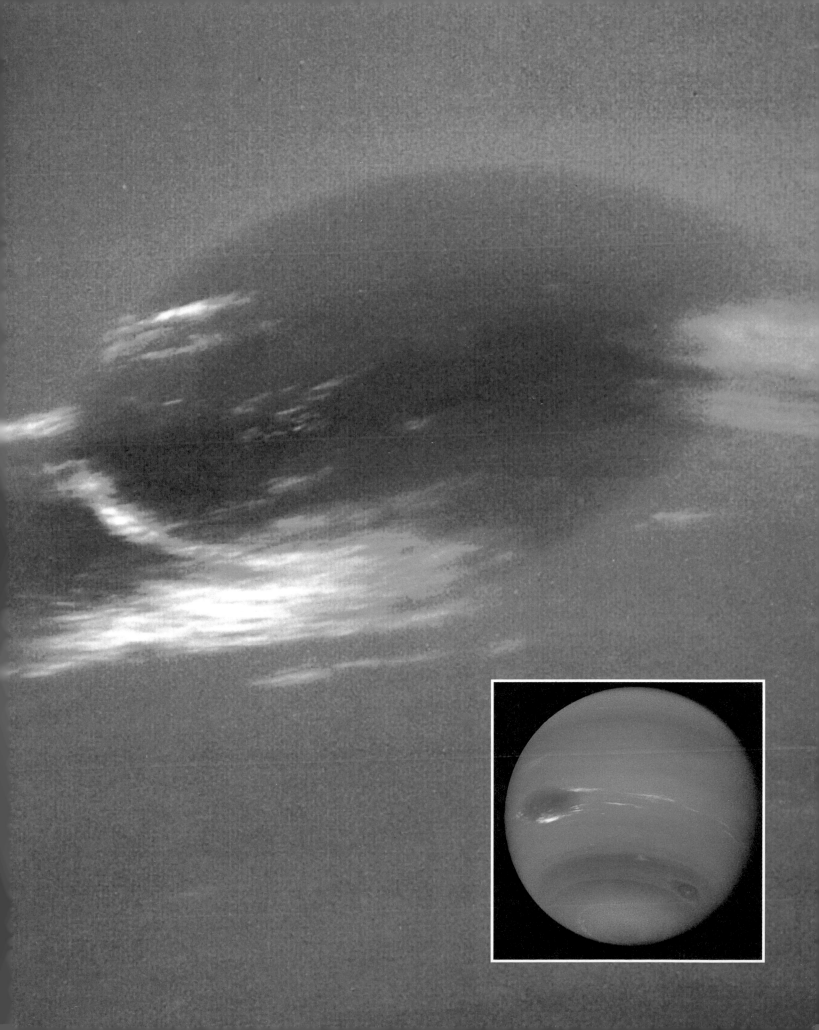

PLUTO

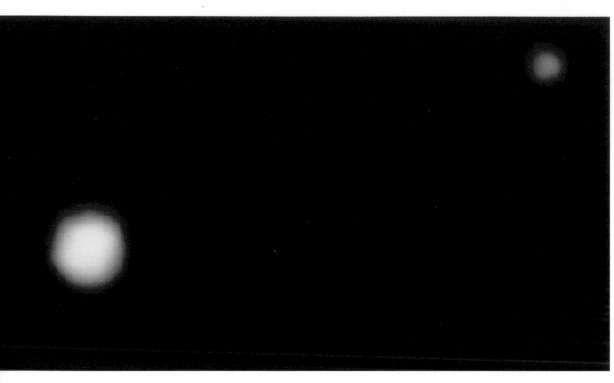

LEFT: The Hubble Space Telescope's best view of Pluto and Charon reveals a planet attended by a moon that is fully half its size. Charon orbits Pluto every 6.4 days, both bodies keeping the same hemisphere facing the other.

The discovery of Pluto in 1930 had been set in motion a number of years before by Percival Lowell (of Martian canal fame, see page 78). An accomplished mathematical astronomer, Lowell believed that a "Planet X" must lie beyond Neptune because both it and Uranus had unexplained irregularities in their motions. (It was discrepancies of a similar kind that had led to the discovery of Neptune.) Lowell calculated positions for Planet X and searched diligently from the observatory he founded in Arizona, but discovered nothing.

FINDING PLANET X

Lowell died in 1916, and his observatory did not strike pay dirt until more than a decade later. Using a new telescopic camera, Clyde Tombaugh eventually came across Planet X on February 18, 1930. It was soon named Pluto, after the Roman god of the underworld. Pluto's orbit is so eccentric that at times it comes closer to the Sun than Neptune does.

While Pluto's orbit was soon computed, the question of its size and mass continued to haunt astronomers. Even large telescopes showed just a faint, starlike point, which suggested it was too small to affect Uranus and Neptune.

LEFT: Clyde Tombaugh was a 22-year-old amateur astronomer and telescope builder from Kansas when he was hired by Lowell Observatory to look for Planet X. He soon found the planet and then extended his search, noting one comet, many asteroids, and several supernovas in distant galaxies.

In 1978 James Christy discovered that Pluto has a moon, now named Charon. Between 1985 and 1990 observers on Earth could see the planet and moon eclipse each other. From this series of eclipses, astronomers were able to

RIGHT AND FAR RIGHT: As no spacecraft has yet visited Pluto, its landscapes remain a mystery. Yet scientists see many parallels with Neptune's moon Triton (right). The two bodies are similar in size and temperature, and Pluto's surface appears to be covered with the same nitrogen and methane frosts as Triton's. Charon's, however, is covered with water frost. At far right is a computer-enhanced image of Pluto taken by the Hubble Space Telescope. The light areas are probably nitrogen or methane frost.

calculate the masses for both objects—and conclude that Tombaugh had been extremely lucky. Pluto has a diameter of about 1,466 miles (2,360 km), while Charon's is half that size at 746 miles (1,200 km). The two bodies rotate, facing one other, on an axis that tilts 122° to Pluto's orbital plane. Together, Pluto and Charon add up to less than ¹⁄₄₀₀ the mass of Earth—which means that they are much too small to affect another planet. Their discovery had been purely fortuitous.

ROCK AND ICE

Pluto's density suggests a rock and ice core covered with layers of ices. Its surface probably resembles Neptune's moon Triton, with nitrogen and methane frosts. The temperature is about −370°F (−220°C).

At present, Pluto is experiencing orbital summer. The evaporating frosts give it an atmosphere of nitrogen and methane, which

its feeble gravity is slowly losing to space. But as Pluto's orbit takes it farther from the Sun, the atmosphere will recondense and fall to the surface as frost.

IS PLUTO REALLY A PLANET?

Pluto fits the classical criteria for a planet: it orbits the Sun, it is large enough to become round, and it even has a moon. But planetary scientists are now starting to think of Pluto (and maybe also Neptune's moon Triton) as objects from the Kuiper Belt. This is a region within the solar system that extends from the zone of the planets out to a thousand astronomical units or farther. Out there, icy planetesimals drift in slow orbits, never becoming sufficiently numerous to assemble into a full-size planet such as Neptune.

Since 1992 astronomers have located more than 60 such objects, of which Pluto and Triton are the largest and probably the most evolved. Using the Hubble Space Telescope, researchers may have detected roughly 40 more objects. Planetary scientists want to study these for the information they may provide about the solar system's earliest days.

Ironically, recent studies have shown that the supposed discrepancies in the motions of Uranus and Neptune resulted from erroneous observations. So Lowell's Planet X was based on faulty data and never existed in reality. Yet it's fortunate that Planet X lived in the minds of astronomers long enough for them to find the real Pluto—at once the last of the Sun's planets and the first of the Kuiper Belt objects.

PLUTO'S ORBIT

Relative to the orbits of all the other planets, Pluto's is more tilted and eccentric. At times this brings the planet closer than Neptune to the Sun.

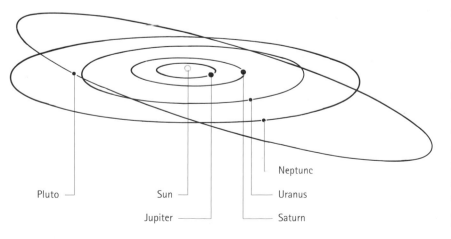

Pluto — Sun — Jupiter — Neptune — Uranus — Saturn

COMETS

Comets are masters of making something out of very little. When a bright comet such as Hyakutake or Hale-Bopp pays us a visit, we delight in its ghostly luminance and huge tail, which may be millions of miles long. Few of us are aware that all the glowing magnificence is provided by the comet's nucleus—a "dirty iceberg" no bigger than a small city.

NUCLEUS AND COMA

A comet is a body of ice, mixed with rock dust and many organic compounds. The only comet yet seen at close range is comet Halley. The *Giotto* spacecraft flew past it in 1986 and photographed its nucleus, an irregular oblong 10 by 5 by 5 miles (16 by 8 by 8 km). The surface of the nucleus is lumpy and cratered, with about the brightness of fresh tar. Its dark crust lets gases escape through its porous, granular structure. Sunlight warms the crust to room temperature or more, while the comet's interior falls to hundreds of degrees below zero. Where the Sun's heat penetrates the crust's thin spots, the ices erupt in jets of dusty gas.

Free of the nucleus, the dust and gas expand into a bright shell called the coma. Water-ice is the main ingredient, but cyanogen, hydrogen cyanide, formaldehyde, methyl alcohol, and methane are also present. Sunlight pushes dust out of the coma, making a yellowish dust tail. The gas in the coma becomes electrically charged by ultraviolet sunlight, and the solar wind shapes it into a bluish gas tail that points away from the Sun like a wind-whipped banner.

FROM BEYOND THE PLANETS

Scientists think comets are the most primitive material left in the solar system. When the solar nebula condensed to form the Sun and planets 4.6 billion years ago, the planetesimals of its inner regions built rocky worlds such as Earth or gas-rich ones such as Jupiter. On the fringes, however, the planetesimals stayed in a kind of deep-freeze and changed little. Today, we call them comets.

Comets lie in two reservoirs beyond the edge of the planets. The comets that pass near Earth come mainly from the first reservoir, the Kuiper Belt, named for Gerard Kuiper, the astronomer who proposed it in 1951. The Kuiper Belt runs from about the orbit of Neptune out to roughly 1,000 astronomical units. It is believed to be flat and disklike, and to merge into the second reservoir, the Oort Cloud, which reaches out to perhaps 100,000 astronomical units, or about 2 light-years. (The Oort Cloud bears the name of Jan Oort, the astronomer who demonstrated its probable existence.) The inner Oort Cloud is a disk like the Kuiper

LEFT: First recorded in Chinese annals in 240 BC, comet Halley last returned to Earth's vicinity in 1986, when a small fleet of spaceprobes sailed out to see it. With a 76-year period, Halley is the best known comet, and has made 30 recorded appearances.

BELOW: Likened to a string of pearls, the separating pieces of comet Shoemaker-Levy 9 numbered more than 20 before striking Jupiter in 1994.

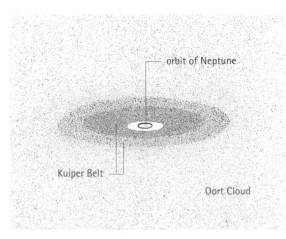

orbit of Neptune

Kuiper Belt

Oort Cloud

THE KUIPER BELT AND THE OORT CLOUD

Beyond Neptune's orbit lies a realm of comets and icy planetesimals. The comets and icy bodies in the Kuiper Belt are what's left of the solar nebula from which all the planets condensed. Farther out, the spherical Oort Cloud is made up of comets that have been flung in all directions by encounters with Jupiter.

Belt, but it flares out to form a large sphere reaching about halfway to the nearest stars.

Scientists have discovered that Neptune is able to grab comets from within the Kuiper Belt and send them inward to Jupiter. Jupiter will either throw them into the inner solar system or fling them out to the Kuiper Belt, the Oort Cloud, or out of the system entirely.

Jupiter can also devour a comet, as the fate of comet Shoemaker-Levy 9 showed in 1994. This hapless comet caught the attention of many people on Earth when Jupiter's gravity broke it into more than 20 pieces. These circled around and slammed into the planet with a force of millions of megatons, leaving dark bruises of dust the size of Earth on the face of Jupiter.

Observing the Sky

We can all delight in the wonders of the night sky just by gazing upward, but greater enjoyment lies in knowing what to look for and how. The possibilities are inspiring. With the naked eye you can chart the wanderings of five of the planets; binoculars reveal Jupiter's four Galilean moons; and through a telescope you can see distant galaxies.

HOW TO BEGIN OBSERVING

There are really two astronomies, not one. So far, this book has dealt with the first kind of astronomy—what you might read about in the newspaper or see reported on TV. This is the astronomy that professional scientists do. It's the fascinating, thrilling, cutting-edge business of pushing back the bounds of the unknown. Professional astronomy may involve measuring the temperature at the Sun's pole, or finding out what lies under the icy crust of Jupiter's moon Europa, or describing the heart of a distant active galaxy that's older than the Sun and Milky Way.

Yet there's another kind of astronomy: the kind you pursue on your own—whether it's from the backyard, a city street, or miles and miles from the nearest streetlights under a black sky powdered with stars. To practice the first kind of astronomy takes long training in physics and mathematics; to engage in the second you simply go outdoors and look up.

GETTING STARTED

This chapter will get you started on the journey. It is divided into four sections, indicated by different colored banding.

The first section (from page 104) begins with some basics about the sky and how it moves, about stars and their brightnesses, and about clocks and time. Then it presents simple star charts for both the Northern and Southern Hemispheres. The charts will help you identify the constellations in their endless flow through the seasons. This is the oldest astronomy of all, since it largely recreates the way that our pretelescopic (and prehistoric) ancestors experienced the sky.

The star charts are followed by the naked-eye section (from page 116), which describes other facets of astronomy without a telescope, and includes a guide to naked-eye sights such as the Moon's phases, solar eclipses, planets, comets, and meteor showers.

The next step involves using optics, which can be as simple as a pair of ordinary binoculars lying around the house. The binocular section (from page 122) describes what to look for when choosing binoculars specifically for astronomy and what you can see with them. The sights described include "seas" of lava on the Moon, beautiful star clusters, and the satellites of Jupiter that Galileo discovered with an instrument less sophisticated than the binoculars of today.

The telescope section (from page 130) concludes the chapter. This section discusses the different kinds of telescopes on the market, how to purchase wisely, and what accessories

COLORED BANDING
This chapter is divided into four sections according to the kind of observing activity described. The colored banding across the top of the page indicates which section that page belongs to.

STAR CHARTS

NAKED–EYE ASTRONOMY

BINOCULAR ASTRONOMY

TELESCOPE ASTRONOMY

LEFT: Comet Hale-Bopp traced a path across the Northern Hemisphere's evening sky in March and April 1997. Easily visible to the unaided eye, bright comets like this were once thought to portend the fates of kings and princes. Today astronomers study comets for what they can tell us about the solar system's history, while skygazers of all kinds enjoy the spectacle of these rare, unpredictable, and very beautiful objects.

RIGHT: *Stars draw curving trails (intersected by a meteor) over one skygazer's observing site. A starry night and a telescope are all you need to explore other planets, colorful double stars, glowing nebulae, and uncountable galaxies.*

you really need to get the most of out of owning a telescope. Then it takes you into the universe with your new instrument to explore such spectacular sights as spots on the Sun, wispy nebulae, and spiral galaxies.

THE LIMITS OF THE UNIVERSE

How far you go in the field of astronomy is mainly determined by your own curiosity. Astronomy—either kind—is an activity for a lifetime of learning and wonder. From here you'll want to explore magazines, astronomy clubs, and the riches of the Internet, which can bring you pictures of new discoveries almost as they are made and link you with other backyard astronomers the world over.

Exploring the heavens—by eye, binocular, or telescope—lets you explore the universe firsthand. You may never travel in space, but anyone can use their binoculars or telescope as a spaceship of the imagination. All it takes on your part is a willingness to step out of the daily routine, a desire to be enthralled, and—most important—a spirit of adventure.

RIGHT: *The Milky Way looms over a field of amateur astronomers, whose red-lensed flashlights scrawl luminous trails in the foreground of this long-exposure photo. In the sky, dark bands of interstellar dust cut across the starclouds that lie in the direction of the Milky Way's center. This is what a spiral galaxy looks like—from the inside.*

MEASURING THE SKY

We watch the Sun disappear below the horizon. The sky darkens and the stars come out. What's happening is that our part of planet Earth is turning eastward, away from its star. The disappearance of the Sun robs the atmosphere of its blue daytime glow and turns it transparent. Our eyes gradually adjust to the darkness, and we look out at the star-filled universe almost unhindered, except for the local glow from city lights.

THE SPHERE OF THE HEAVENS

The sky poses an illusion. In reality it extends infinitely far, filled with stars of all types, brightnesses, and distances. But to our earthbound eyes, it appears as a hemisphere covering us like a dome. This impression is so strong that we can adopt it as a convenient fiction, calling it the celestial sphere to include that part hidden by the solid Earth underfoot.

Earth spins on an axis, and the points on the celestial sphere directly above the North and South Poles are called the celestial poles. Halfway between the poles on Earth is the Equator, which lies directly below a corresponding line in the sky called the celestial equator.

SKY LATITUDE AND LONGITUDE

Astronomers locate objects in the sky with coordinates similar to latitude and longitude. As latitude measures distance north or south of the Equator, declination measures angular distance from the celestial equator. It runs from 0° at the equator to 90° (north or south) at the

ABOVE RIGHT: The imaginary celestial sphere has been employed for centuries. This 17th-century version shows "The Situation of Earth in the Heavens."

THE CELESTIAL SPHERE

The celestial sphere's westward motion (arrow) reflects Earth's eastward rotation. Because Earth's axis tilts relative to its orbit, the plane of its orbit—the ecliptic—meets the celestial equator at an angle.

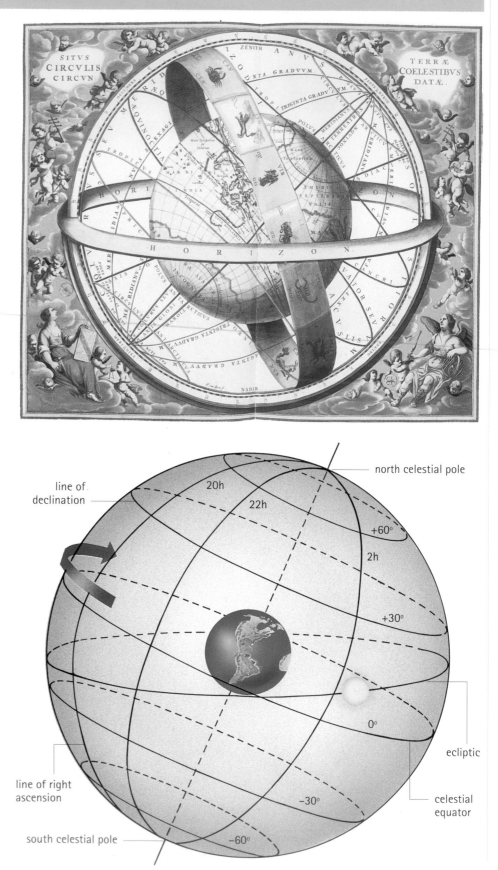

celestial poles, and is measured in degrees, minutes, and seconds of arc. (These units are also used to indicate the apparent size of a celestial object—that is, how big it appears to be in our sky—and the apparent separation of objects—that is, how far apart two objects appear to be.)

The other coordinate on Earth is called longitude. In the sky, this coordinate is called right ascension, and it is measured eastward from the point where the Sun stands on the first day of spring in the Northern Hemisphere. But unlike the units of arc used for declination, right ascension's units are hours, minutes, and seconds of time. This reflects the fact that the biggest clock in the world is the world itself.

TIMEKEEPING

Knowing when a sky event occurs is difficult because of Earth's many time zones. Astronomers use Universal Time (UT), which is essentially Greenwich Mean Time. In North America, UT runs five hours ahead of Eastern Standard Time or eight hours ahead of Pacific Standard Time. (Adjust the figures to four or seven hours during daylight time.)

Conversion is easy: if a lunar eclipse reaches maximum at 4h45m UT on January 21st, the local time in New York is 11:45 PM on January 20th, while in Los Angeles it is 8:45 PM, also on the 20th.

BIG AND BRIGHT

Stars vary enormously in distance and luminosity. Together these two factors determine how bright a star appears in the sky. For more than 2,000 years astronomy has used magnitudes to describe star brightness. The brightest star in the sky, Sirius in Canis Major, shines at −1.5 magnitude, while the faintest stars visible to the eye are about 5.0, or 5th, magnitude. The scale is such that a 1st magnitude star is 100 times brighter than a 5th magnitude star, and negative magnitudes indicate brighter stars than do positive magnitudes.

To compare the true luminosity of different stars, astronomers use a measurement called absolute magnitude—how bright a star would be at a distance of 32.6 light-years. On this scale, the Sun comes out pretty puny. In our sky it's at −26.8 magnitude, but its absolute magnitude is only +4.8. On the other hand, the absolute magnitude of Sirius is +1.4. Thus, Sirius is 3.4 magnitudes, or 23 times, more luminous than the Sun.

ABOVE: Betelgeuse, the orange star in the upper left of Orion, appears nearly the same brightness as Rigel, the blue-white star on the lower right. Yet Betelgeuse lies nearly three times farther from Earth than Rigel does. They look about the same only because Betelgeuse has a much greater absolute magnitude.

RIGHT: To gain a rough idea of apparent distances, hold your hand at arm's length: with fingers spread, your hand covers about 20 degrees of sky; a fist covers about 10 degrees; and a thumb, 2 degrees.

HOW TO READ A STAR CHART

The night sky can be very confusing, and it's easy to feel lost in space. For most people, the sky is unfamiliar territory where nothing carries a signpost. On the other hand, there's no need to hurry when learning your way around the stars.

FINDING YOUR DIRECTION

The first step in using a star chart is to find north (or south, if you live in the Southern Hemisphere). In the Northern Hemisphere, the bowl of the Big Dipper in Ursa Major makes it easy to locate Polaris, the star lying nearest the pole.

For observers in the Southern Hemisphere, the Southern Cross acts as a pointer, but not so accurately, and the South Pole lacks a bright star near it. (See diagrams at right.)

USING THE CONSTELLATION CHARTS

With north and south identified, turn to the appropriate constellation chart for the current season (pages 108 to 115). The charts show only the brightest stars, and will be easiest to use at first if you are under a dark sky but not a completely black, rural one. If the sky is extremely dark, you'll see so many stars that they will overwhelm the basic constellations.

Stars differ in apparent brightness, so the charts rank them by graduated dots. The lines on the charts connect the brighter stars into patterns called constellations. These had mythological meaning to past civilizations and now serve to divide up the sky. There are 88 official constellations, but since stars lie at all different distances, the patterns have little physical reality.

Pick either the north- or south-looking chart and read the description. (A small flashlight with red paper over the lens lets you read the charts without ruining night vision.) Usually, there's one bright constellation or star you can identify easily. Take your time and locate it in the sky. Remember to mentally adjust the scale of the chart to the real sky—each chart shows the whole horizon between

east and west, and runs from the horizon to straight overhead. After locating one star pattern, use it to jump off to others.

AS THE WORLD TURNS

If you stay outside for an hour, you will notice the stars have shifted westward. This apparent movement is caused by Earth's rotation. Stars at the western edge of the map slip out of sight, while new stars appear in the east. If you look toward the celestial pole, you'll see that prominent constellations have turned like the hands of a big clock. Constellations near the pole don't ever set; they pass below it and begin to rise again. Such stars are called circumpolar.

Constellations also drift westward as the calendar unrolls, which reflects Earth's yearly progress in orbit around the Sun. When our part of Earth turns away from the Sun and night falls each evening, the starry backdrop that we look out on is slightly different. Every night, any given star will rise about four minutes earlier because of this slow movement. Four

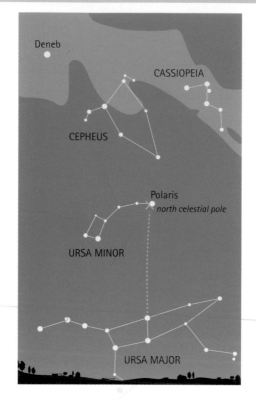

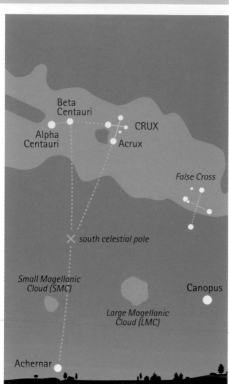

ABOVE: Both celestial poles have "pointers," but the north is easier to find because the Big Dipper in Ursa Major is familiar and the star Polaris lies near the pole. In the south, Crux makes the best guide.

ABOVE: A really dark sky can make constellations harder to find. The Big Dipper stands on its handle (right) while its bowl stars point to Polaris (left).

minutes a day isn't much, but it adds up. After a month's time, it means that a given star or constellation will rise two hours earlier (or set two hours sooner).

THE WALTZ OF THE PLANETS

As you explore constellations, you'll see bright stars that don't appear on the maps. Most will shine steadily instead of twinkling. These interloper "stars" are actually planets, a word derived from the Greek for "wanderer." (Don't mistake airplanes or satellites for planets—both will move visibly in a few seconds, whereas planets won't.) If you watch the sky carefully over many weeks you will notice that planets move relative to the stars around them—this is why they were called planets. To know which planets are visible and when, check the sky calendars published in astronomy magazines.

USING A STAR ATLAS

Once the major constellations become familiar, you should invest in a star atlas— a bound set of sky charts. They show more stars and other objects than are shown on the charts in this book. An atlas that goes down to about 7th magnitude is good for binoculars and small telescopes. With a larger telescope, get one that goes fainter.

When you first see a map in a star atlas, it appears bewildering. Star atlases once carried pretty drawings of mythical figures; today you get thousands of dots, mysterious labels, lines you can't see in the sky—and even east and west are reversed compared to an ordinary map.

Take your time and study the charts. All star atlases plot lines of right ascension and declination. They label stars with their names or other designations such as numbers or Greek letters. Special symbols mark double and variable stars, while star clusters, nebulae, and galaxies have their own symbols and M or NGC numbers.

Star atlases may look intimidating at first, but they contain a lot of order—and they are your roadmap to the universe.

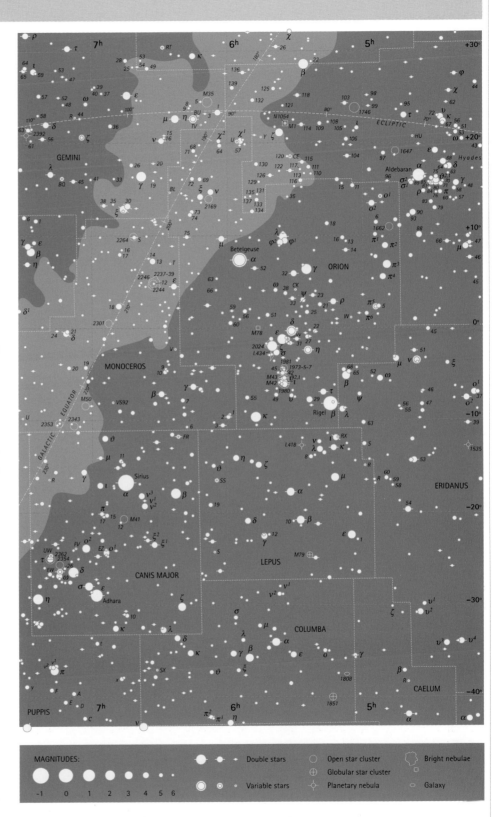

ABOVE. This star-atlas map shows the familiar figure of Orion, the Hunter, and the constellations around it, plus their official boundaries. It depicts stars to 6th magnitude and plots many non-stellar objects such as nebulae and clusters. Part of the Milky Way shows in light blue along the left edge.

MAGNITUDES:
-1 0 1 2 3 4 5 6

Double stars
Variable stars
Open star cluster
Globular star cluster
Planetary nebula
Bright nebulae
Galaxy

NORTHERN HEMISPHERE STAR CHARTS

SPRING

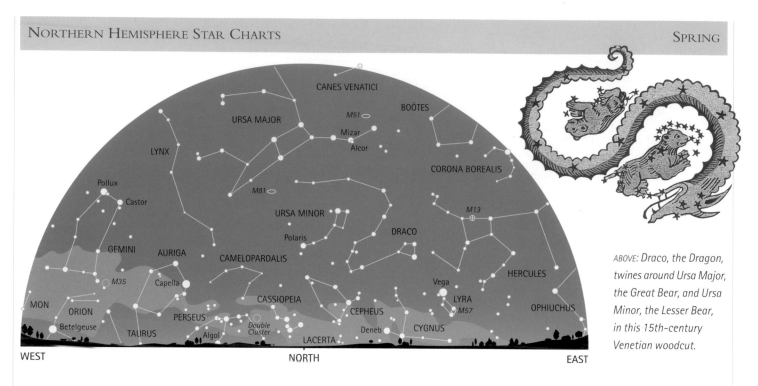

ABOVE: *Draco, the Dragon, twines around Ursa Major, the Great Bear, and Ursa Minor, the Lesser Bear, in this 15th-century Venetian woodcut.*

LOOKING NORTH The landmark of the northern sky is Ursa Major, the Great Bear. It lies tonight above the pole star, Polaris, which is in Ursa Minor, the Lesser Bear. Ursa Major's brightest stars form the famous Big Dipper. The two stars that form the lefthand side of the Dipper's bowl point down to Polaris. Standing in the northwest are Gemini, the Twins, with Castor and Pollux. In the northeast, bright Vega heralds the rise of Lyra, the Lyre, with Hercules above. Follow the curving handle of the Dipper as it "arcs to Arcturus" in Boötes, the Herdsman, who sits high in the southeast.

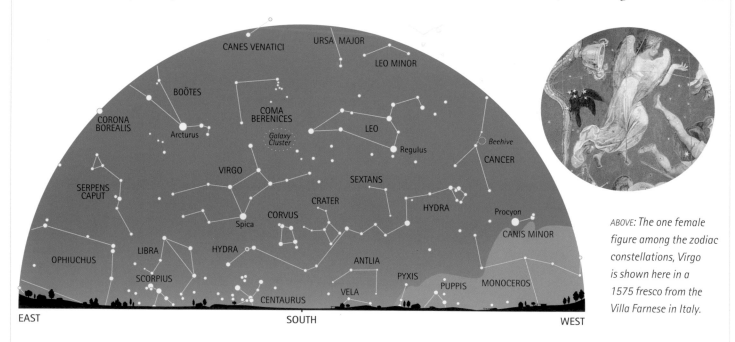

ABOVE: *The one female figure among the zodiac constellations, Virgo is shown here in a 1575 fresco from the Villa Farnese in Italy.*

LOOKING SOUTH As Procyon in Canis Minor, the Lesser Dog, edges westward, three bright stars light the southern sky— Arcturus in Boötes, the Herdsman (in the southeast), Spica in Virgo, the Maiden (south), and Regulus in Leo, the Lion (southwest). Below Leo and Virgo runs the night sky's longest constellation— Hydra, the Water Snake—but it is hard to see except on dark and moonless nights. Between Virgo and the tail of Leo lies the Realm of the Galaxies, the nearest cluster of galaxies to our own. This group of spiral and elliptical galaxies lies about 65 million light-years away.

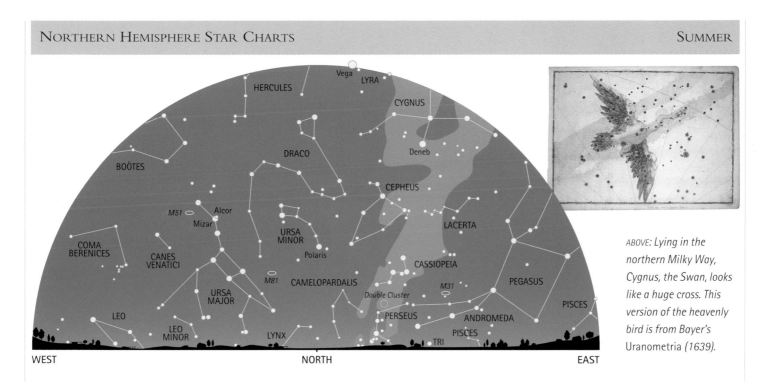

ABOVE: Lying in the northern Milky Way, Cygnus, the Swan, looks like a huge cross. This version of the heavenly bird is from Bayer's Uranometria (1639).

LOOKING NORTH The Big Dipper (in Ursa Major, the Great Bear) lies to the left of the pole star, Polaris, opposite the W-shape of Cassiopeia, the Queen, rising on the right. Above Ursa Minor, the Lesser Bear, look for the dim form of Draco, the Dragon, which wraps around the pole. If the night is moonless and dark, you can see the Milky Way span the sky from northeast to southwest, passing from Perseus, the Hero, through Cassiopeia and Cepheus, the King, to Cygnus, the Swan. Cygnus is also known as the Northern Cross. It appears as a bird flying south along the Milky Way.

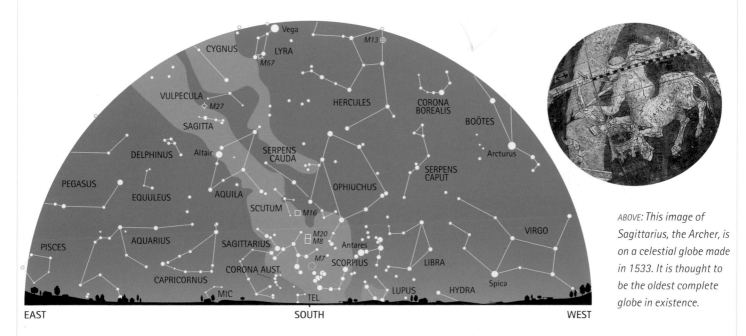

ABOVE: This image of Sagittarius, the Archer, is on a celestial globe made in 1533. It is thought to be the oldest complete globe in existence.

LOOKING SOUTH The richest part of the Milky Way parades across the south on summer nights, featuring the constellations of Scorpius, the Scorpion, and Sagittarius, the Archer. (Sagittarius resembles the profile of a teapot, with Milky Way "steam" rising from its spout.) Here lies the mysterious center of the Milky Way Galaxy, hidden behind 25,000 light-years of dusty gas. Farther north, the Milky Way divides alongside Aquila, the Eagle, and Cygnus, the Swan, because a cloud of interstellar dust near the Sun is obscuring the distant stars. You will see it best on a moonless night away from city lights.

NORTHERN HEMISPHERE STAR CHARTS

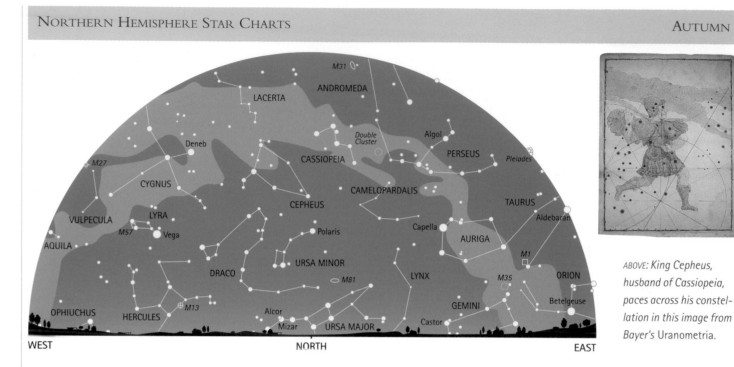

ABOVE: *King Cepheus, husband of Cassiopeia, paces across his constellation in this image from Bayer's* Uranometria.

LOOKING NORTH The Milky Way spans the sky for those viewing away from streetlights on a moonless night. To the north, the bright stars of the Big Dipper in Ursa Major, the Great Bear, graze the horizon, while Cassiopeia, the Queen, makes a bent M-shape above Polaris, the pole star. As Auriga, the Charioteer, rises in the northeast, the Summer Triangle of the stars Vega, Deneb, and Altair is descending in the west. High overhead lies the great Andromeda Galaxy (M31)—the closest major galaxy to us. It is the most distant object visible to the naked eye, at 2.2 million light-years away.

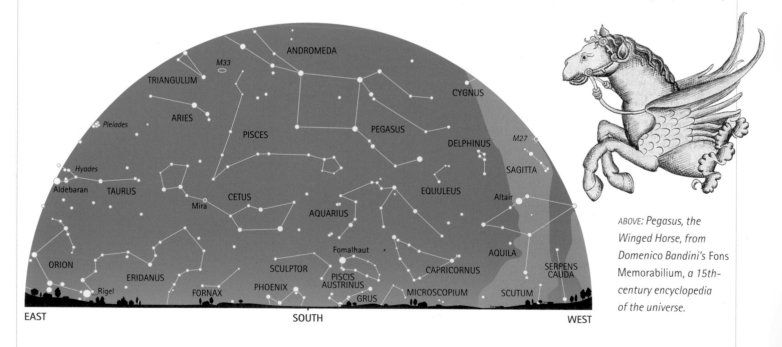

ABOVE: *Pegasus, the Winged Horse, from Domenico Bandini's* Fons Memorabilium, *a 15th-century encyclopedia of the universe.*

LOOKING SOUTH High in the south, the Great Square of Pegasus is a notable landmark. The Flying Horse is upside down, so the line of stars to the lower right of the Square delineates his neck. This line reaches west toward the Milky Way and the bright star Altair in Aquila. Look below the Great Square to spot various "watery" constellations: Pisces, the Fishes; Cetus, the Whale; Aquarius, the Water Carrier; and Piscis Austrinus, the Southern Fish. In the eastern sky, Taurus, the Bull, is making his appearance, followed by winter's Orion, the Hunter, still partially hidden below the horizon.

NORTHERN HEMISPHERE STAR CHARTS WINTER

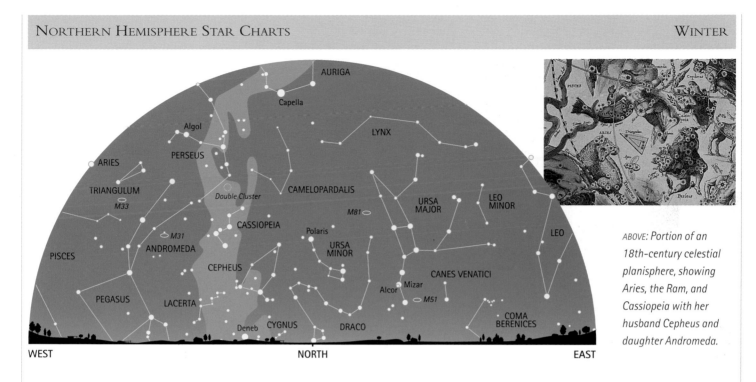

ABOVE: Portion of an 18th-century celestial planisphere, showing Aries, the Ram, and Cassiopeia with her husband Cepheus and daughter Andromeda.

LOOKING NORTH The Big Dipper in Ursa Major, the Great Bear, stands on its handle, while the two top bowl stars point left to Polaris in Ursa Minor, the Lesser Bear. As the Dipper rises, Cassiopeia, the Queen, sinks, along with her mythological companions—Cepheus (her husband), Andromeda (her daughter), Perseus (Andromeda's hero and savior), and Pegasus (Perseus' horse). Appropriately, Cetus, the Whale, who was about to devour Andromeda when Perseus intervened, has already set. In the east, Leo, the Lion, is rising, marked by the 1st magnitude star Regulus.

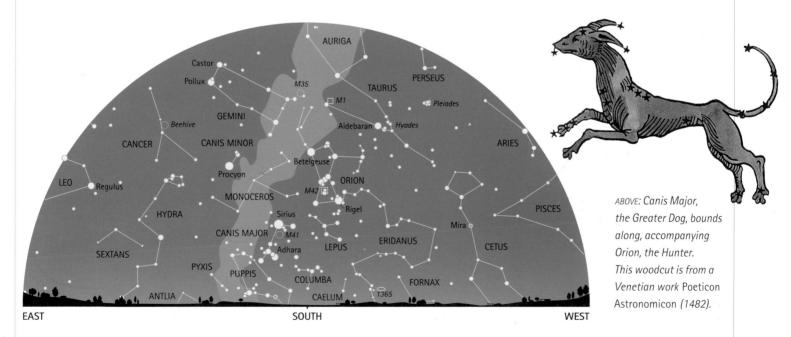

ABOVE: Canis Major, the Greater Dog, bounds along, accompanying Orion, the Hunter. This woodcut is from a Venetian work Poeticon Astronomicon *(1482).*

LOOKING SOUTH Orion, the Hunter, strides across the horizon, driving back Taurus, the Bull, with the lovely Pleiades star cluster on its back. Taurus' face is formed from a looser star cluster, the Hyades. To the lower left of Orion, Canis Major, the Greater Dog, follows the Hunter, with Sirius (the brightest star in the sky) marking his eye. Left of Orion, the star Procyon distinguishes Canis Minor, the Lesser Dog, while Castor and Pollux stand at the head of Gemini, the Twins. Alongside Gemini is Auriga, the Charioteer, wheeling overhead along the Milky Way.

Southern Hemisphere Star Charts   Spring

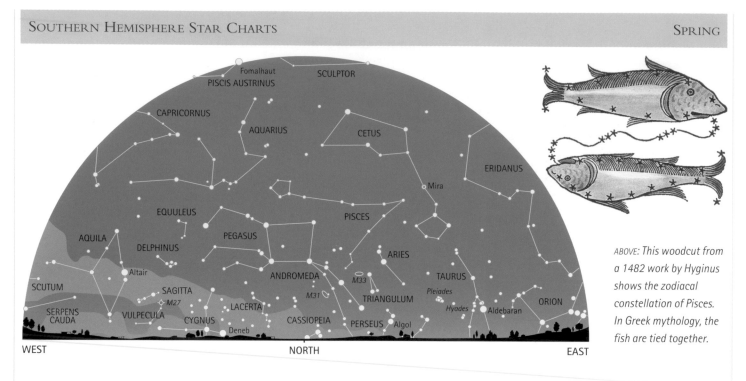

WEST — NORTH — EAST

ABOVE: *This woodcut from a 1482 work by Hyginus shows the zodiacal constellation of Pisces. In Greek mythology, the fish are tied together.*

Looking North The Great Square of Pegasus, the Flying Horse, forms a landmark as he gallops west across the northern horizon. The two lower stars in the Square point west toward bright Altair in Aquila, the Eagle. The two stars on the left side of the Square point up to Fomalhaut in Piscis Austrinus, the Southern Fish, high overhead. Between Pegasus and Fomalhaut lie many dim "watery" constellations: Pisces, the Fish; Cetus, the Whale; and Aquarius, the Water Carrier. Rising in the east is Aldebaran—the ruddy eye of Taurus, the Bull—with Orion, the Hunter, on the horizon.

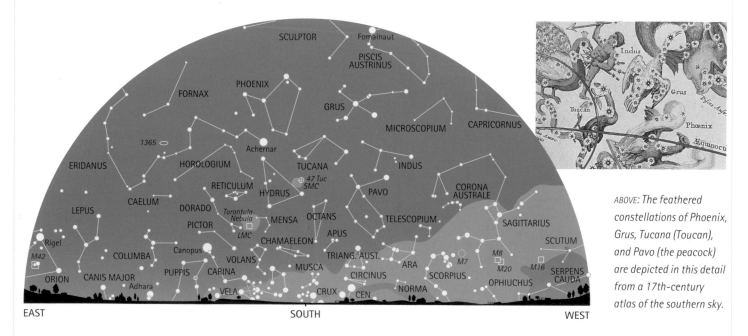

EAST — SOUTH — WEST

ABOVE: *The feathered constellations of Phoenix, Grus, Tucana (Toucan), and Pavo (the peacock) are depicted in this detail from a 17th-century atlas of the southern sky.*

Looking South Two bright stars make a triangle with the south celestial pole: Achernar in Eridanus, the River, and Canopus in Carina, the Keel. The pole lies about where the lower right star would be. Between Canopus and Achernar lies the Large Magellanic Cloud (LMC), a satellite galaxy of the Milky Way, with the Small Magellanic Cloud (SMC) slightly above it. Several celestial birds flock here: Phoenix, the Firebird, above Achernar; Grus, the Crane, to Phoenix' right; and Tucana, the Toucan, and Pavo, the Peacock, below them both. The Milky Way lines the western horizon.

SOUTHERN HEMISPHERE STAR CHARTS   SUMMER

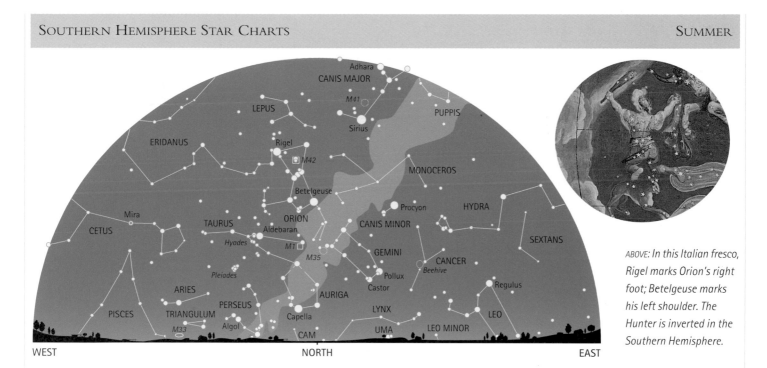

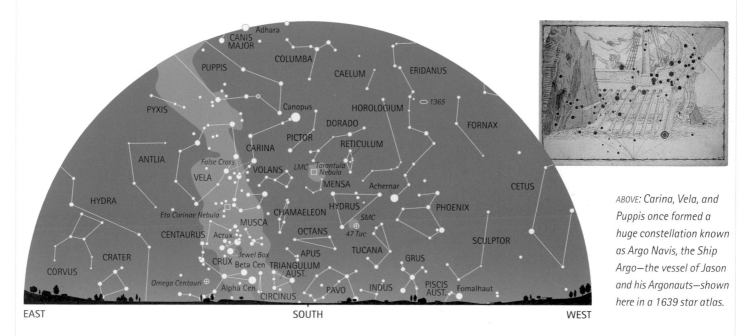

ABOVE: In this Italian fresco, Rigel marks Orion's right foot; Betelgeuse marks his left shoulder. The Hunter is inverted in the Southern Hemisphere.

LOOKING NORTH The tall figure of Orion, the Hunter, with his three-starred Belt, stands high in the north. Extending the Belt down to the left points toward Taurus, the Bull, with ruddy Aldebaran and the Hyades and Pleiades. Extending the Belt upward to the right points to Sirius, the sky's brightest star, in Canis Major, the Greater Dog. From Sirius, a line down to the northeast passes Procyon in Canis Minor, the Lesser Dog, to Regulus in Leo, the Lion. Below Orion, yellow Capella arcs low with Auriga, the Charioteer. To its right, stand the bright Twins of Gemini—Castor and Pollux.

ABOVE: Carina, Vela, and Puppis once formed a huge constellation known as Argo Navis, the Ship Argo—the vessel of Jason and his Argonauts—shown here in a 1639 star atlas.

LOOKING SOUTH High in the southern sky the bright star Canopus is the rudder in Carina, the Keel. The rest of the ship is made up by Vela, the Sails, and Puppis, the Stern, both of which lie in the Milky Way. Rising in the southeast below Vela is tiny Crux, the Southern Cross, with Centaurus, the Centaur, wrapped around it. During this season, the southwestern sky has a single bright star, Achernar, a beacon at the end of Eridanus, the River. Look between Achernar and Canopus for a misty patch—the Large Magellanic Cloud. It forms a triangle with the Small Magellanic Cloud and Achernar.

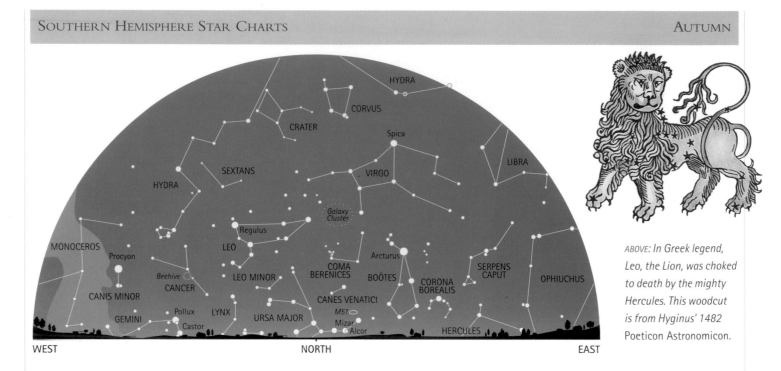

WEST   NORTH   EAST

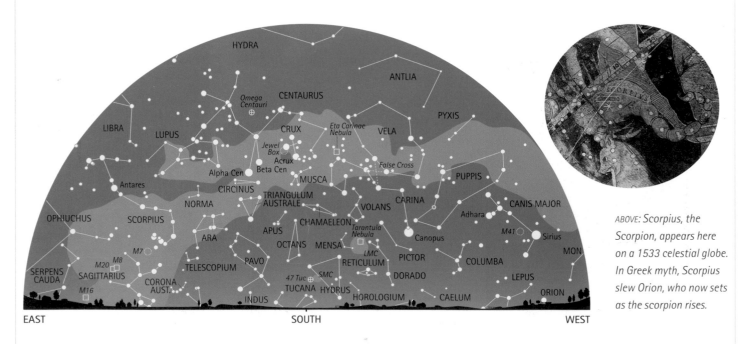

ABOVE: *In Greek legend, Leo, the Lion, was choked to death by the mighty Hercules. This woodcut is from Hyginus' 1482 Poeticon Astronomicon.*

LOOKING NORTH Four bright stars make it easy to find constellations this season. In the northwest, look for Procyon in Canis Minor, the Lesser Dog. In the north, spot Regulus in Leo, the Lion. Then in the northeast, warm-tinted Arcturus stands in Boötes, the Herdsman, with white Spica in Virgo, the Maiden, above. Between Leo and Virgo lies the Virgo cluster—the nearest cluster of galaxies to our own, at about 65 million light-years away. High overhead, the long figure of Hydra, the Water Snake, has his head close to Procyon while his body weaves past Crater, the Cup, and Corvus, the Crow.

EAST   SOUTH   WEST

ABOVE: *Scorpius, the Scorpion, appears here on a 1533 celestial globe. In Greek myth, Scorpius slew Orion, who now sets as the scorpion rises.*

LOOKING SOUTH The glories of the Milky Way arch across the south during spring, from Sirius in Canis Major, the Greater Dog, setting in the west, to ruddy Antares in Scorpius, the Scorpion, rising in the southeast. Above Canopus in the southwest lies Carina, the Keel; Vela, the Sails; and Puppis, the Stern—all part of Jason's mythical ship *Argo*. To the left of these stands tiny Crux, the Southern Cross, which Centaurus, the Centaur, so nimbly hops over, his forefeet marked by Alpha and Beta Centauri. If possible, get away from city lights and explore the Milky Way with a pair of binoculars.

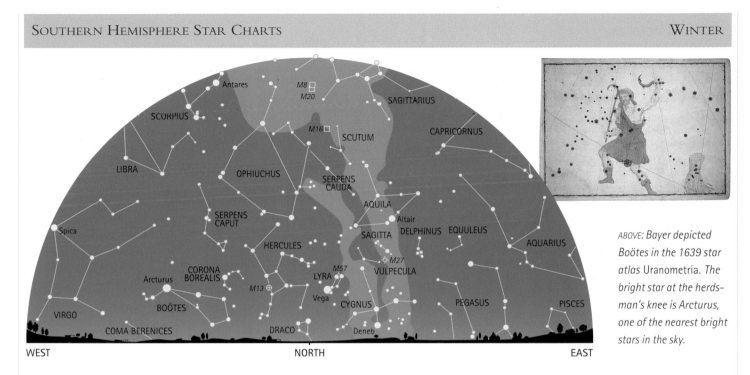

ABOVE: Bayer depicted Boötes in the 1639 star atlas Uranometria. The bright star at the herdsman's knee is Arcturus, one of the nearest bright stars in the sky.

LOOKING NORTH Three bright stars line the northern horizon. In the northwest, Arcturus sets with Boötes, the Herdsman. Due north stands Vega, hallmark of Lyra, the Lyre; higher in the northeast, in the Milky Way, is Altair in Aquila, the Eagle. (Beneath them, you may spot Deneb in Cygnus very low in the sky.) Right of Boötes lie Corona Borealis, the Northern Crown, and Hercules. If the night is dark and moonless, look for the great rift in the Milky Way near Aquila, caused by a huge cloud of dust. The Milky Way widens overhead because there lies the populous heart of our galaxy.

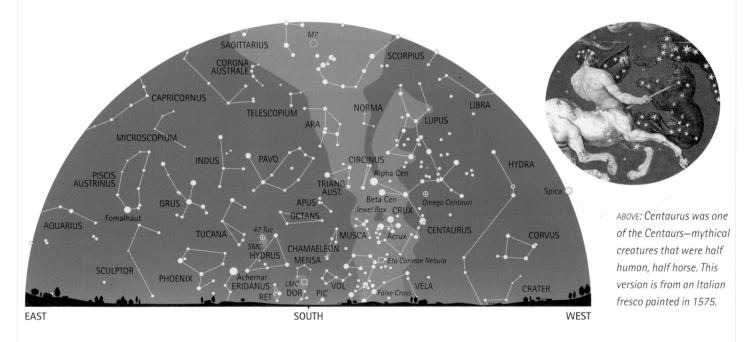

ABOVE: Centaurus was one of the Centaurs—mythical creatures that were half human, half horse. This version is from an Italian fresco painted in 1575.

LOOKING SOUTH The Milky Way is setting in the southwest and taking with it some of the brightest stars. Crux, the Southern Cross, is easy to identify, while Centaurus straddles it. Alpha and Beta Centauri mark the Centaur's forelegs. Above the Centaur creeps Lupus, the Wolf, and a star-rich run of Milky Way that reaches to Scorpius, the Scorpion, high overhead. Here, and in adjoining Sagittarius, the Archer, lie many of the Milky Way's greatest sights. On the next moonless night, find an observing site away from city lights and explore this region with binoculars or a telescope.

NAKED-EYE ASTRONOMY

Observing the sky with the unaided eye is the simplest kind of astronomy. It puts you in touch with the earliest sky-gazers, those first humans who looked up and tried to understand the lights in the night sky. Beyond the simple pleasures of learning the constellations, naked-eye astronomy offers many remarkable sights. Some of these, such as five of the planets or the phases of the Moon, can be seen on almost any night just by walking out the door. Others, such as meteor showers and solar eclipses, are events that can be predicted and warrant some preparation.

Modern civilization, with its glow of artificial lights, puts many barriers in the way of appreciating the sky. Yet star-strewn heavens are not necessarily lost forever, just a bit beyond one's everyday reach. If you have ever marveled at the beautiful Milky Way during vacation, recapturing that feeling at other times of year simply takes a little planning.

BELOW: The glow of lights over cities has quietly stolen the natural sky, and compels skywatchers to seek observing sites in more remote areas.

OUTFITTING FOR SPACE

Many preparations for observing hold whether you're using the unaided eye or binoculars or a telescope. The priorities are: stay warm and comfortable, find a dark-sky site that's convenient to get to, and bring this book and perhaps a star atlas. You might also want to pack a camera and tripod (see below).

Overdressing is better than underdoing it, because even summer nights can be cool. Bring bug repellent if the season or location calls for it. And a thermos of hot chocolate, coffee, or tea will taste very good after an hour or two outside.

To explore the skies, you will want to use the star charts in this book or a star atlas, but remember that reading these by white light will ruin your night vision for a while, so pack a red-filtered flashlight.

YOUR OBSERVING LOCATION

When looking for a good observing site, use a map to pick an area with as rural a surrounding as you can get to in about half an hour's drive. Your goal is to leave as many streetlights behind as you can. Find a dark side road with little traffic and a clear, unobstructed horizon. Make sure you can pull over and park on the shoulder where cars will not come on you abruptly.

As your eyes adjust to the dark, locate north or south and look for the brighter stars. Soon you'll see many more and the constellations will fall into place.

BASIC SKYSHOOTING

It's easy to capture the naked-eye sky on film. Detailed photos of the Moon or galaxies require a telescope and considerable experience, but a simple camera can take impressive sky scenes. Here's how to get started.

Choose a camera that can take exposures of an hour or longer. (This rules out many point-and-shoot models.) You'll need a tripod and fast film (at least ISO 400). Pick slide film for the first few rolls, so you can see exactly what the camera records. Wide-angle lenses will produce dramatic views, but even a normal lens works well, as do telephotos.

Set the camera on the tripod, focus the lens to infinity, open its aperture fully, and aim toward a prominent constellation. Start with an exposure of 1 minute, then roughly double the time with each frame: 1 minute, then 2, 5, 10, 15, and 30 minutes, and finally 1 hour. (Jot down how long each frame runs.) Snap an ordinary daylight scene or two at the start and end of the roll so the processor cuts the frames correctly. (Or ask that the film be left unmounted.)

Compare the finished slides with the exposure record to see what worked best. Short exposures capture stars and constellations much as you see them. But a sky filled with curving star trails (and perhaps a meteor) is quite dramatic too, especially when the camera points north and the trails curl around Polaris.

RIGHT: Star-trail photos work best when skies are dark enough to permit exposures lasting an hour or more. Try moonless nights at a rural site.

DESCRIPTION Following the lunar phases is one of the oldest pleasures in astronomy. After New Moon, look for earthshine—sunlight reflecting off Earth that illuminates the portion of Moon not lit by the Sun. At Full Moon, a lunar eclipse may occur.

VISIBILITY Between New Moon and Full, the Moon rises during daytime and is most easily visible in the evening sky. Between Full and New, it rises after sunset and is seen best between midnight and dawn.

APPARENT MAGNITUDE Ranges from −8.8 (crescent) to −12.7 (Full Moon)

APPARENT SIZE 32 arcminutes

ACTUAL DIAMETER 2,160 miles (3,476 km)

ORBITAL PERIOD 29.5 days from one New Moon to the next, and 27.3 days relative to the stars

DISTANCE 238,856 miles (384,401 km)

LEFT: The Full Moon rising, partly in eclipse.

DESCRIPTION In a total solar eclipse, the Moon blocks the disk of the Sun and we can see the soft "flower" of the corona, the Sun's hot outer atmosphere. Even the partial phases look dramatic as the Moon's disk slowly covers more and more of the Sun. However, observers must take precautions to prevent blindness (see page 74).

VISIBILITY A total solar eclipse occurs somewhere in the world about once a year. While the total eclipse can be seen from only a small area, a partial eclipse is visible over a much larger region.

APPARENT MAGNITUDE OF SUN −26.7

ABSOLUTE MAGNITUDE OF SUN 4.8

APPARENT SIZE OF SUN 32 arcminutes

ACTUAL DIAMETER OF SUN 865,278 miles (1,392,530 km)

DISTANCE TO SUN 1 AU, or 93 million miles (150 million km)

LEFT: The Sun in partial eclipse.

DESCRIPTION To the naked eye, a planet looks generally starlike but does not twinkle as stars do. A second characteristic is that over a few days or weeks, you can see a planet change position relative to the real stars. At first glance, planets appear colorless. But a more careful look reveals Mars' warm ocher hue, which contrasts with Jupiter's icy white gleam. Saturn looks ivory-white, while Mercury often has a warm tinge, though not as pronounced as Mars'. Brilliant white Venus can even cast a dim shadow.

VISIBILITY Planets are generally visible every night in the year, although not all are visible at once or on any given date. Astronomy magazines and some newspapers describe which planets are visible at what times. The unaided eye can easily see the five planets known since antiquity. Mercury and Venus are visible for a time after sunset or before sunrise, while Mars, Jupiter, and Saturn may be seen at any time of night, depending on where they are in the sky.

APPARENT MAGNITUDE This varies as the distances from Earth and the Sun change.

Mercury: −2.0 to +5.0

Venus: −4.0 to −4.6

Mars: −1.3 to +1.2

Jupiter: −2.0 to −2.5

Saturn: 0.0 to +1.0

APPARENT SIZE Apparent size varies with distance from Earth.

Mercury: 5 to 12 arcseconds

Venus: 10 to 64 arcseconds

Mars: 4 to 25 arcseconds

Jupiter: 32 to 49 arcseconds

Saturn: 16 to 20 arcseconds; with rings, 36 to 46 arcseconds

DISTANCE This varies during the orbit.

Mercury: 0.6 to 1.4 AU

Venus: 0.3 to 1.7 AU

Mars: 0.5 to 2.5 AU

Jupiter: 4.2 to 6.2 AU

Saturn: 8.5 to 10.5 AU

RIGHT: *The crescent Moon with Venus above, Jupiter below, and faint Mercury among the tree branches.*

NAKED-EYE SIGHTS — AURORAS

DESCRIPTION Auroras take many forms: shimmering curtains of color, pulsating arcs, and flickering shafts of light are the most common and striking. The displays usually appear light green, with reddish tints indicating more energetic activity.

VISIBILITY Auroral activity waxes and wanes over the 11-year sunspot cycle and often follows outbursts on the Sun (see page 58). While auroras have been seen in all parts of the world, they are most common at high latitudes. Displays are often limited to polar regions of the sky: in the Northern Hemisphere, the northern sky is the best place to look, while the southern sky is better in the Southern Hemisphere. Unfortunately moonlight and especially city lights cover up all but the strongest displays.

DISTANCE 60 to 100 miles (100 to 160 km)

LEFT: An auroral arc over Alaska.

NAKED-EYE SIGHTS — COMETS

DESCRIPTION The word *comet* is derived from the Greek for "hairy star," which does describe how a comet looks. Comets typically display a straight bluish tail of ionized gas and a curving whitish tail of sunlit dust (see page 98).

VISIBILITY Comets arrive unpredictably. Apparitions typically run several weeks, with best visibility lasting about a week.

APPARENT MAGNITUDE May range from −3 (or brighter) to barely visible

APPARENT SIZE Comet tails can stretch across more than 100 degrees of sky, but most measure 30 to 50 degrees.

ACTUAL SIZE The tail of a comet can be millions of miles long, while the coma, or head, may extend many thousands of miles. Comet nuclei range from ½ mile (1 km) to 30 miles (50 km) in diameter.

DISTANCE 0.015 AU to 2 light-years

LEFT: Comet Hale-Bopp above Zion Canyon, Utah.

METEOR SHOWERS

DESCRIPTION In a meteor shower, bright streaks of light radiate from the constellation that carries the shower's name. Also, on any night you'll see a few "sporadic" meteors per hour, strays unassociated with any known shower. Many meteors look yellowish or green.

VISIBILITY Meteor showers hit peak activity on a particular date, but reduced numbers can be seen a day or two before and after. The greatest number of meteors is visible in the after-midnight hours. Moonlight may wash out fainter meteors.

MAJOR SHOWERS Quadrantids (January 3), Lyrids (April 22), Eta Aquarids (May 5), Delta Aquarids (July 28), Perseids (August 12), Orionids (October 22), Leonids (November 18), Geminids (December 14)

DISTANCE 50 to 75 miles (80 to 120 km)

RIGHT: The straight trail of a meteor cuts across the curved star trails in this long-exposure photograph.

NAKED-EYE SIGHTS   ARTIFICIAL SATELLITES

DESCRIPTION The easiest way to spot an artificial satellite is to lie back and look steadily at one part of the sky. Be alert for a star that is noticeably moving. Many satellites change brightness as they turn or tumble in sunlight. To distinguish aircraft from satellites, look for the strobe light or red and green navigation lights that all aircraft carry but satellites don't. (Binoculars can confirm a naked-eye identification.)

VISIBILITY Satellites are visible every clear night of the year, although the best times to look are in dark twilight (evening or morning)—when the sky appears black but an object orbiting some 100 miles (160 km) up is in full sunlight.

APPARENT MAGNITUDE Range from −4 to barely visible

DISTANCE 100 to 22,300 miles (160 to 35,900 km)

RIGHT: A satellite trail approaches comet Austin.

SELECTING BINOCULARS

Binoculars make an ideal first "telescope" for several reasons. First is price. In the United States, good astronomy binoculars start at around $200, substantially less than a decent telescope costs. Second is ease of operation. Binoculars have wider fields of view, and they also give upright images (whereas most telescopes produce inverted images)—both features make locating celestial objects easier. The third virtue is portability. Telescopes have mountings that must be erected for use; binoculars are pretty much grab-and-go.

While any binoculars will do to start with, when purchasing binoculars for astronomy keep the following in mind.

READING THE SPECS

Binoculars are identified by a number such as 7x50. The "7x" indicates the magnification, and the "50" gives the main lens diameter—or aperture—in millimeters. When observing celestial objects, light grasp is important, so skywatchers generally opt for binoculars with 50 mm or larger lenses. Another useful number is the exit pupil, found by dividing the first number into the second. A pair of 7x50 binoculars will produce an exit pupil of 7.1 mm. The exit pupil should be as big as the pupil of your eye when it is fully dilated in the dark. For young people this figure is about 7 mm, but as you age past your twenties, it drops closer to 5 mm.

FIELD OF VIEW

If you wear glasses, look for long "eye relief." This determines how close your eye must be to the eyepiece to see the whole field of view. With some binoculars, the field of view is marked in degrees. If not, they give the field as, for example, 375 feet at 1,000 yards, or 130 meters at 1,000 meters. Divide the field by 52.4 (feet at 1,000 yards) or by 17.5 (meters at 1,000 meters) to get the field in degrees. Thus, a field of 375 feet at 1,000 yards equals a field of about 7 degrees.

Field of view depends partly on magnification, and wider is better, all in all. What's best? Most observers feel that below 7x you don't get as striking a view, and few people can hold binoculars steady at more than 10x. In most binoculars, powers between 7x and 10x offer at least a 5-degree field of view, often larger.

TWO BINOCULAR DESIGNS

Binoculars are two low-power telescopes aligned exactly parallel. Roof-prism models (top) tend to be more compact and much more expensive than porro-prism ones (below). But porros work perfectly well for astronomy and will give many years of pleasure.

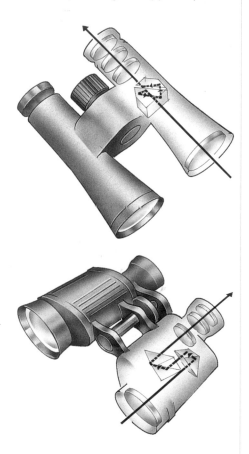

BELOW: The Pleiades star cluster in Taurus—as seen in 7x50 binoculars with about a 7-degree field of view (left), and in 20x80s with 3.5-degree field.

ABOVE AND BELOW: Two models of 7x50 porro-prism binoculars. When you buy from a store, you can easily check the binoculars for smoothness of focus, clear views right to the edge of the field, and overall fit to your hands and eyes.

BUYING WISELY

Binoculars come in two designs, porro prism and roof prism. Porros have a "dogleg," while roof-prism models look like the letter H. Optically, neither design offers advantages, although porro prisms tend to cost less. Roof-prism models are usually sealed and waterproof, which pushes up the price—but since little astronomy is done in the rain, this feature matters more to birders and hunters.

The better 7x50s and most larger models have threaded sockets that attach to a camera tripod, and tripod-adapters are available for many porro models. Tripods help with steadiness, but aren't much use for scanning directly overhead. So take a lawn chair, put the back down, and brace your elbows on the armrests.

Buying binoculars by mail order can save some money, but you lose the chance to try out models. Everyone's hands differ slightly, and fit is important. Plus it is all but impossible to check eye relief without actually handling the instrument.

There are some binoculars that are definitely not appropriate for astronomy. Avoid zoom binoculars—they have narrow fields of view and poor optical quality. Likewise, shun any "fixed focus" or "focus-free" models—you need to be able to adjust the focus to suit your eyes.

REALLY BIG BINOCULARS

Resembling giant 7x50s, big binoculars carry specifications like 11x70, 20x80—even 25x100. These make a poor first choice for beginners, because they are expensive—costing as much as a small telescope—and are very difficult to hand-hold. But if you have been exploring astronomy for a while, consider them as a second pair. They are fun to use and can give eye-filling views of the sky.

ABOVE: Binoculars are not just for beginners. Many experienced skywatchers use them for scanning wide fields of sky easily and quickly.

BELOW: Tripods are essential with binoculars as big as these 11x70s. But even smaller models give better views when you brace them or use a tripod.

DESCRIPTION The face of the "Man in the Moon" is drawn by the lunar "seas," or maria. They generally appear dark gray (like weathered asphalt), but binoculars reveal variations in tint, which indicate differences in composition, age, and texture. The circular maria fill impact basins—the ancient scars of giant meteorite hits. Rays—bright, whitish streaks around the younger craters—are splashes of rock powder thrown out by meteorite impacts.

VISIBILITY The maria are visible at any phase. Rays stand out best at Full Moon.

ACTUAL SIZE Mare Serenitatis, a circular maria, is about 370 miles (600 km) in diameter. The ray from Tycho that crosses Mare Serenitatis lies more than a thousand miles from its source.

DISTANCE 238,856 miles (384,401 km)

LEFT: The bright ray-streaked southern highlands and large seas of dark lava dominate the Full Moon.

DESCRIPTION A pair of binoculars lets you experience what Galileo saw when he discovered Jupiter's four largest moons—Io, Europa, Ganymede, and Callisto. Each evening at the same time, look for tiny points of light flanking Jupiter and draw what you see. Within a month the pattern of the moons' movements will be clear.

VISIBILITY Any or all may be visible on a given night, depending on where they are in their orbit around Jupiter.

APPARENT MAGNITUDE Io +5.0, Europa +5.3, Ganymede +4.6, Callisto +5.7

ACTUAL DIAMETER Io 2,256 miles (3,630 km), Europa 1,950 miles (3,138 km), Ganymede 3,270 miles (5,262 km), Callisto 2,983 miles (4,800 km)

ORBITAL PERIOD Io 1.77 days, Europa 3.55 d, Ganymede 7.15 d, Callisto 16.69 d

DISTANCE 4.2 to 6.2 AU

LEFT: The four Galilean moons flank Jupiter.

BINOCULAR SIGHTS

MIZAR AND ALCOR

DESCRIPTION The stars Mizar and Alcor are about 30 light-years apart, but they appear close together in our skies, forming an optical double star. They can be split by eye, but are best seen with binoculars. Mizar is a true double star, but you will need a telescope to see its two components.

VISIBILITY The pair is visible all year long for Northern Hemisphere viewers outside of the tropics, and for at least part of the year from the tropics of both hemispheres.

CONSTELLATION Ursa Major

APPARENT MAGNITUDE Mizar 2.4, Alcor 4.0

ABSOLUTE MAGNITUDE Mizar 2.3, Alcor 1.8

APPARENT SEPARATION Mizar and Alcor appear to be 12 arcminutes apart.

DISTANCE Mizar about 60 light-years, Alcor about 90 light-years

RIGHT: The optical double star Mizar and Alcor.

BINOCULAR SIGHTS

THE PLEIADES (M45)

DESCRIPTION This open star cluster is popularly known as the "Seven Sisters." Most people can detect only six Pleiades by eye, but binoculars reveal many more. Astronomers estimate the Pleiades contain roughly 500 stars scattered across 100 light-years of space. A large telescope will reveal the wispy cloud of starlit dust that envelops the stars, but the cluster itself is best seen through the wide field of view provided by binoculars.

VISIBILITY The Pleiades can be seen from everywhere except Antarctica.

CONSTELLATION Taurus

APPARENT MAGNITUDE 1.2

APPARENT SIZE The brightest stars of the cluster form a tiny dipper-shape about 1 degree across.

ACTUAL DIAMETER 100 light-years

DISTANCE 360 light-years

RIGHT: The Pleiades open star cluster.

BINOCULAR SIGHTS   DOUBLE CLUSTER (NGC 869 AND 884)

DESCRIPTION These two open star clusters contain more than 100 stars each, and lie about 100 light-years apart in space. One of the two looks slightly more concentrated in brightness toward its center, and both have what appear as strings of stars leading out of them. Astronomers estimate that the clusters are about 3 to 5 million years old.

VISIBILITY On a dark, moonless night, the pair is visible to the naked eye as an oval hazy spot in the Milky Way halfway between Perseus and Cassiopeia. The Double Cluster never rises for latitudes south of the southern tropics.

CONSTELLATION Perseus
APPARENT MAGNITUDE +3.5 (each)
APPARENT SIZE 30 arcminutes (each)
ACTUAL SIZE 65 light-years (each)
DISTANCE About 7,300 light-years

LEFT: The Double Cluster in Perseus.

BINOCULAR SIGHTS   JEWEL BOX CLUSTER (NGC 4755)

DESCRIPTION This compact open cluster benefits from the 10x to 20x magnification of large binoculars. It shows as a bright grouping of stars, many of them displaying tints that echo and contrast with its brightest star, 6th magnitude Kappa Crucis. At about 7 million years old, the Jewel Box is relatively young.

VISIBILITY The cluster lies in a rich part of the southern Milky Way, adjacent to the dark cloud of dust known as the Coal Sack. In most locations, the Jewel Box is easily spotted by the naked eye on a clear night, but it never rises north of the northern tropics.

CONSTELLATION Crux
APPARENT MAGNITUDE +4.2
APPARENT SIZE 10 arcminutes
ACTUAL SIZE 22 light-years
DISTANCE Roughly 7,600 light-years

LEFT: The contrasting tints of the Jewel Box's stars.

BINOCULAR SIGHTS   ORION NEBULA (M42)

DESCRIPTION The Great Nebula in Orion is an object worth examining in any size instrument from binoculars to telescope. To the naked eye, it is a dim patch of haze in Orion's Sword; in binoculars, it glows bluish green. (A telescope reveals greater detail.) The nebula is a dense region of dusty gas where new stars are forming, as witnessed by the Hubble Space Telescope (see page 56). The light of multitudes of hot new stars causes the nebula's gas to glow.

VISIBILITY The Orion Nebula is visible from everywhere in the world except areas within a few degrees of the North Pole.

CONSTELLATION Orion

APPARENT MAGNITUDE +2.9

APPARENT SIZE 1 degree

ACTUAL SIZE 25 light-years across

DISTANCE 1,500 light-years

RIGHT: The Orion Nebula in the Hunter's Sword.

BINOCULAR SIGHTS   HERCULES CLUSTER (M13)

DESCRIPTION Like all globulars, the Hercules Cluster appears as a round ball of fuzz. (Novice comet-hunters often mistake globulars for comets.) Binoculars reveal its brighter center, but you will need a medium-sized telescope to resolve individual stars. Edmond Halley (of comet fame) discovered this cluster in 1714.

VISIBILITY The cluster sits on one side of the main figure of Hercules. It can be seen from everywhere in the world except the southern high latitudes. On moonless nights away from city lights, the cluster is just visible to the naked eye.

CONSTELLATION Hercules

APPARENT MAGNITUDE +5.9

APPARENT SIZE 17 arcminutes

ACTUAL SIZE 112 light-years across

DISTANCE 22,500 light-years

RIGHT: The globular cluster M13 in Hercules packs about 300,000 stars into a tight ball.

BINOCULAR SIGHTS   ANDROMEDA GALAXY (M31)

DESCRIPTION This spiral galaxy, the closest large galaxy to the Milky Way, is a beautiful hazy oval visible to the naked eye. With binoculars, it appears slightly enlarged and shows a brighter central region, its nucleus. If skies are very dark and you are using large binoculars, you may catch a hint of the dark dust lanes that define Andromeda's spiral arms. The galaxy belongs to the Local Group, a cluster of more than 30 galaxies, including the Pinwheel Galaxy (see below).

VISIBILITY It is visible from north of the Southern Hemisphere's tropics.

CONSTELLATION Andromeda

APPARENT MAGNITUDE +3.4

APPARENT SIZE 3 degrees x 1 degree

ACTUAL SIZE 124,000 light-years

DISTANCE 2.2 million light-years

LEFT: The spiral Andromeda Galaxy is the most distant object the naked eye can see.

BINOCULAR SIGHTS   PINWHEEL GALAXY (M33)

DESCRIPTION Because it spreads over a large area, the Pinwheel Galaxy is better viewed in the wide field of binoculars than through a telescope. It is notoriously tricky to spot—you'll need a very dark night and a viewing site well away from any city lights. The galaxy shows as a large oval and may be easier to locate if you look slightly to one side and catch it out of the corner of your eye. Seen through binoculars, the nucleus and some of the disk appear as a soft gray glow.

VISIBILITY It can be seen from anywhere north of the high southern latitudes.

CONSTELLATION Triangulum

APPARENT MAGNITUDE +5.7

APPARENT SIZE 40 arcminutes x 1 degree

ACTUAL SIZE 52,000 light-years

DISTANCE 2.9 million light-years

LEFT: The Pinwheel is the third largest galaxy in the Local Group, after the Milky Way and M31.

BINOCULAR SIGHTS   LAGOON NEBULA (M8)

DESCRIPTION Lying in a star-filled region of the Milky Way, this emission nebula is dimly visible to the naked eye on a dark moonless night. Binoculars show it as a fuzzy oval patch of light north of the Teapot of Sagittarius. If you observe carefully, you may detect the dark lane of dust that cuts across the nebula's middle and looks something like a lagoon. Like all emission nebulae, the Lagoon is a cloud of dust and hydrogen gas made to glow by the hot stars within and around it (see pages 38–39).

VISIBILITY It is visible from everywhere except the northern high latitudes.

CONSTELLATION Sagittarius

APPARENT MAGNITUDE +4.6

APPARENT SIZE 40 x 90 arcminutes

ACTUAL SIZE 60 x 140 light-years

DISTANCE 5,200 light-years

RIGHT: A dark dust lane bisects the Lagoon Nebula.

BINOCULAR SIGHTS   LARGE MAGELLANIC CLOUD (LMC)

DESCRIPTION Looking like a detached piece of the Milky Way, the Large Magellanic Cloud is easy to spot by eye, even on a night with a bright Moon. It is a wonderful object to explore in any size binoculars (or telescope), being rich in stars, clusters, and nebulae. An irregular satellite galaxy of our own Milky Way, both it and the Small Magellanic Cloud are believed to be remnants of a larger population; the other members have collided and merged with the Milky Way.

VISIBILITY The galaxy can be seen from everywhere south of the northern tropics.

CONSTELLATIONS Dorado and Mensa

APPARENT MAGNITUDE +0.1

APPARENT SIZE 9 x 11 degrees

ACTUAL SIZE 31,000 light-years

DISTANCE 168,000 light-years

RIGHT: The Large Magellanic Cloud is classed as an irregular galaxy, but shows signs of a central bar.

SELECTING A TELESCOPE

Among backyard astronomers a telescope is almost a badge of membership. So it's tempting to run out and buy something with a white tube and a few eyepieces. However, the wiser course is to get a pair of binoculars first (see pages 122–123) and start setting money aside. In the United States, for instance, a few good telescopes sell for less than $500, but $1,000 is a more realistic minimum. This may seem like a lot to outlay, but it is in line with other hobbies such as photography. A telescope good enough to support a lasting interest is not a mass-market item and usually involves skilled hand-finishing. Yet it will last essentially forever since there are few parts to wear out or break in use.

HOW BIG?

Telescopes are ranked not by how much they magnify, but by their aperture—the diameter of their main lens or mirror. The rule of thumb is: the larger the aperture, the more you'll see. But size imposes a price in portability; ask yourself how easy it will be to carry the telescope into the backyard—and to transport it by car out to dark skies. A small scope that's used often beats a giant that collects dust.

TYPES OF TELESCOPE

Telescopes fall into three basic kinds: refractor, reflector, and catadioptric.

Refractors use a lens to collect light and focus it. In sizes of 3.1 or 3.5 inches (80 or 90 mm), they are solid yet portable and easy to use. Avoid 2.4 inch (60 mm) refractors—most of these are poor quality.

Reflectors gather and focus light with a curved mirror. Money buys the most aperture with this kind of telescope, although they require occasional adjustment to keep the optics aligned. A 4 to 6 inch (100 to 150 mm) makes a good beginner's scope. A Newtonian reflector on a Dobsonian mounting (see facing page) yields even more aperture for the price, and sizes up to 10 inches (250 mm) are affordable as starter instruments.

Catadioptric telescopes use a mirror plus a lens to focus the light. The most popular models are the 8 inch (200 mm) size. While considerably more expensive, they combine compact size with substantial aperture in an easy-to-transport package.

No telescope design rates best for all purposes—and all are versatile enough to provide satisfying views of celestial objects.

TAKING AIM

Telescopes use one of two mountings: altitude-azimuth or equatorial.

The simplest is the altitude-azimuth, often abbreviated to alt-az. These swing up and down and pivot left and right. To follow celestial objects the mounting uses both movements together. Alt-az mountings may be moved by hand or by motor under computer control. The Dobsonian mounting is an alt-az mount using simple, low-cost materials.

The other form of mounting is the equatorial. These mountings also have two axes, but one points to the celestial pole. This design lets the mounting track the sky with a single movement. Larger, heavier, and more expensive than most alt-az mounts, equatorials are also powered by hand or by electric motor.

refractor

finderscope

eyepiece

focusing controls

German-style equatorial mounting

computer-control handset

LEFT: This 5 inch (125 mm) refractor rides on a German-style equatorial mounting. The polar axis aims upward at the celestial pole and a computerized motor drives the telescope to counteract the westward movement of the sky. Many computer-driven telescopes have databases of objects and can locate them on demand.

BEING A SAVVY BUYER

A telescope costs serious money, so research accordingly. Pore over the reviews in astronomy magazines, check their special annual issues for surveys of telescopes, read the ads, and send away for catalogs. Also, join a local astronomy club and gather advice from other members. Attend at least one star party and look through various makes of telescope. Studying telescopes for a year before buying is not too long to wait.

When you do buy, pay extra for a sturdier mounting. You won't regret it. Consider purchasing the next-larger size of mounting than the one your telescope would usually come with. Check for stability by focusing on a distant object at about 150x magnification and tapping the tube. If the vibrations take several seconds to die away, the telescope will drive you crazy because every time you adjust the focus—or the wind blows—the image will dance and wobble.

When shopping for a telescope, try to visit a range of specialist stores. Avoid department and discount stores, and never buy a scope marketed by its magnification with claims such as "See 1,000x!!" Almost without exception, these are junk.

ABOVE: Prices may be a little higher at a telescope store, but you will find amateur astronomers on staff who can help you make a wise selection.

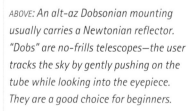

ABOVE: An alt-az Dobsonian mounting usually carries a Newtonian reflector. "Dobs" are no-frills telescopes—the user tracks the sky by gently pushing on the tube while looking into the eyepiece. They are a good choice for beginners.

RIGHT: This Schmidt-Cassegrain telescope is on an equatorial fork mounting. The fork arms point to the celestial pole and a computer-controlled motor keeps the telescope aimed at the stars as Earth rotates.

THREE TELESCOPE DESIGNS

Telescopes collect light in three main ways. Reflectors (top) use a mirror, which reflects light to an eyepiece at the front of the telescope. Refractors (middle) use a refracting lens and focus the light at the bottom of the scope. Catadioptric telescopes (bottom), which include Schmidt-Cassegrains, combine a correcting lens with a series of mirrors, and have the eyepiece at the back of the tube.

SELECTING ACCESSORIES

Accessories are more than just "frosting" on the telescope-cake. They improve and augment your telescope's capabilities, and they let you undertake observations that you couldn't otherwise make. Advertised through astronomy magazines and manufacturers' catalogs, accessories are available in all price ranges and levels of quality.

EYEPIECES

The most important telescope accessory is a good eyepiece. The job of a telescope's main lens or mirror is to form an image of a celestial object, and the eyepiece precisely magnifies the image so you can see it clearly. Most telescopes are sold with at least one eyepiece, but marketing pressures keep the quality fairly low. It is wise to plan to upgrade the eyepieces soon after buying a new telescope. Luckily you don't need many.

Two quantities, its own focal length and that of the telescope, govern an eyepiece's power—its ability to magnify. Thus, one eyepiece used on two different telescopes will yield two different powers. The power of an eyepiece is found by dividing its focal length into that of your telescope. Eyepiece focal lengths are given in millimeters, but telescopes are usually specified by focal ratio (abbreviated as f/ratio) and aperture. To get the focal length, multiply the aperture by the f/ratio, converting to millimeters if necessary. An 8 inch f/10 telescope, for example, has a focal length of 80 inches, or 2,032 mm. On such a telescope a 28 mm eyepiece gives a power of 73x—that is, it would magnify the image 73 times.

Backyard observing often uses relatively low magnifications. Because Earth's atmosphere is seldom very still, few nights permit using powers like 500x or 1,000x. Most observers find that two or three eyepieces handle 90 percent of their observing. Tastes vary, but three quality eyepieces giving about 75x, 150x, and 300x will cover most situations. In the United States, quality eyepieces start at about $50 each. High-quality eyepieces (which can cost up to several hundred dollars each) offer sharper, more contrasty images with wider fields of view.

UPSIDE DOWN

One eyepiece that you don't need is a terrestrial-viewing one. Telescopes produce an inverted image, which skywatchers quickly become used to. A terrestrial-viewing eyepiece has additional optics to turn the image upright again. This may be useful on Earth, but for celestial viewing that extra glass simply wastes light—of which there is little enough already.

ABOVE: Useful accessories include eyepieces (back), glass filters (center), a Barlow lens to double the power of any eyepiece used with it (front), and an adapter to attach a camera to the telescope (right). Buy the best quality accessories you can afford.

RIGHT: Two views of the Eagle Nebula in Serpens. A finderscope shows about 5 degrees of sky (left), while a telescope shows about 1 degree (right). The finderscope's wide field of view makes it ideal for locating objects too faint to see by eye.

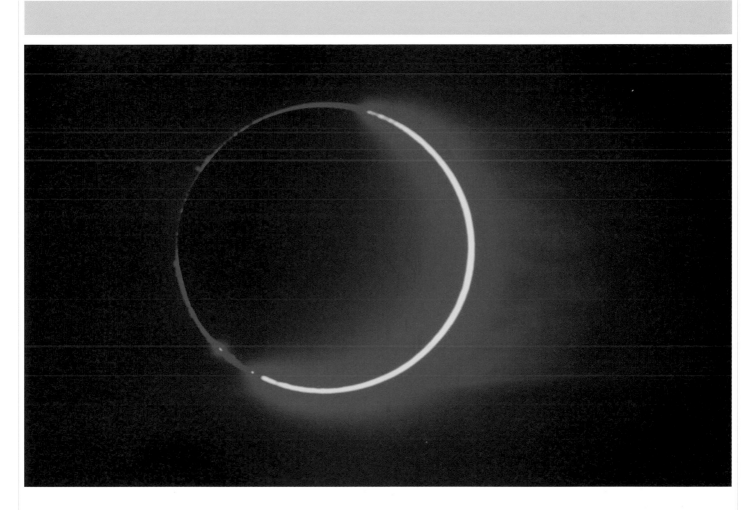

One helpful eyepiece accessory is a star diagonal. This bends the optical path by 90°, letting you view objects high in the sky without straining your neck. Star diagonals are most useful with refractors and catadioptric telescopes.

FINDERSCOPES

After eyepieces, the next most important accessory is a good finderscope. This is a wide-field, low-power telescope that is aligned parallel to the main telescope's axis and helps you center on an object. (Most finders have crosshairs to make aiming a cinch.) Finderscopes are speci-fied like binoculars (see page 122) and bridge the gap between the naked-eye view of the sky and that given by your lowest-power eyepiece. While telescopes are usually supplied with a finder, it's often just a 6x30 or smaller. Experienced observers soon replace this with an 8x50.

FILTERING THE VIEW

Filters are another useful accessory—and essential for solar observing. The only safe solar filters are those designed to fit over the aperture of the telescope. These are made of aluminized Mylar or glass. (Avoid those designed to fit over the eyepiece—they can crack suddenly and blind you without warning.)

Other filters can make glowing nebulae or planetary details stand out more clearly. With planets, details are often subtle and low in contrast, and colored filters can help you see them. A red filter, for example, can make the dark markings on Mars more prominent, while a blue filter will enhance any Martian clouds present. A light blue filter works with Jupiter.

Filters that reject the glow from streetlights and other human-caused light pollution will increase the visibility of nebulae. These filters, often called light-pollution-reduction (or LPR) filters, cost more than ordinary colored filters, but offer real improvements, especially for observers in city locations.

LEFT AND ABOVE: Colored planet filters screw into the eyepiece, but a safe solar filter attaches to the front of the telescope and must entirely cover the aperture. An aperture filter lets you safely observe and photograph the stages of a solar eclipse (above).

TELESCOPE SIGHTS   MARTIAN POLAR CAPS

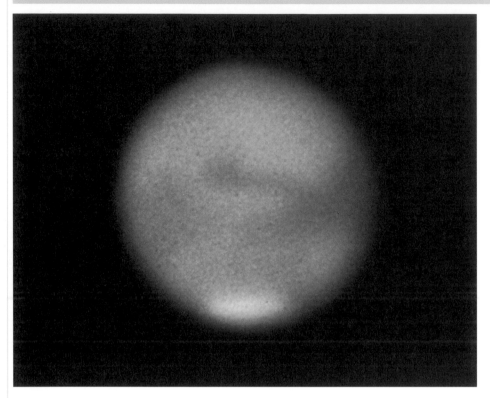

DESCRIPTION Through a telescope at high power (150x or more) when the view is calm and steady, Mars' ice caps appear as whitish patches at the northern and southern poles. A red filter accentuates the polar caps, while a blue or violet filter enhances the hood of cloud that often forms over whichever polar region is experiencing local winter. The southern cap varies more in size from winter to summer because the south pole undergoes greater temperature extremes.

VISIBILITY At least one cap can be seen whenever Mars appears. Whether both caps show depends on the Martian season.

APPARENT SIZE OF MARS Ranges from 4 to 25 arcseconds

ACTUAL DIAMETER OF MARS 4,217 miles (6,787 km)

DISTANCE 0.5 to 2.5 AU

LEFT: The southern polar cap on the red ball of Mars.

TELESCOPE SIGHTS   SATURN'S RINGS

DESCRIPTION The rings become visible in a telescope at about 20x magnification, but if the view is steady, use 100x or more for a clear image. In a backyard telescope, three rings are visible. From outward in, they are the A, B, and C rings. The dark Cassini Division separates the A ring from the B ring. The C ring appears as a gray band.

VISIBILITY The rings are usually visible from Earth whenever Saturn is. Until 2002 the rings' southern side opens wider to view; afterward the rings narrow until 2009, when they are edge-on to us and disappear briefly from view.

APPARENT SIZE 36 to 46 arcseconds

ACTUAL SIZE OF A RING 85,000 miles (137,000 km)

DISTANCE 8.5 to 10.5 AU

LEFT: The ringed planet is probably the most eye-catching sight of all in backyard astronomy.

TELESCOPE SIGHTS — SUNSPOTS

DESCRIPTION Through a properly protected telescope (see page 133), sunspots appear as dark markings on the Sun's face. At low power (50x), they look like black spots, but at higher powers, they display more structure: a grayish "collar" (the penumbra) surrounds a dark core (the umbra). Sunspots form where intense magnetic activity prevents energy from deeper levels reaching the surface. The spots are cooler than their surroundings.

VISIBILITY The number of sunspots varies over an 11-year cycle. They can be seen on almost any clear day, except during periods of minimum activity.

APPARENT SIZE Up to 1 arcminute

ACTUAL SIZE Up to 30,000 miles (50,000 km)

DISTANCE 1 AU, or 93 million miles (150 million km)

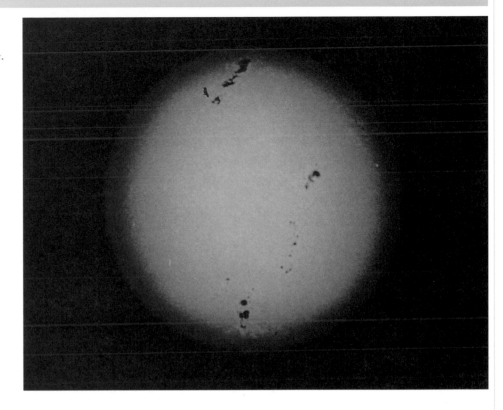

RIGHT: Clusters of spots on the surface of the Sun.

TELESCOPE SIGHTS — RING NEBULA (M57)

DESCRIPTION In a small telescope at 50x, the Ring appears as a tiny gray patch, slightly oval in outline. (The colors in photos are invisible to the eye.) With a medium to large backyard telescope and more power, the darker center becomes visible and the nebula resembles a ghostly smoke ring. It takes a large instrument (16 inch, or 400 mm) to pick out the 15th magnitude central star. The Ring is a planetary nebula made from the outer layers of this star, which were thrown off as the star evolved (see page 38).

VISIBILITY It can be seen from anywhere north of the southern middle latitudes.

CONSTELLATION Lyra

APPARENT MAGNITUDE +8.8

APPARENT SIZE 60 x 80 arcseconds

ACTUAL SIZE 0.4 x 0.5 light-year

DISTANCE 1,300 light-years

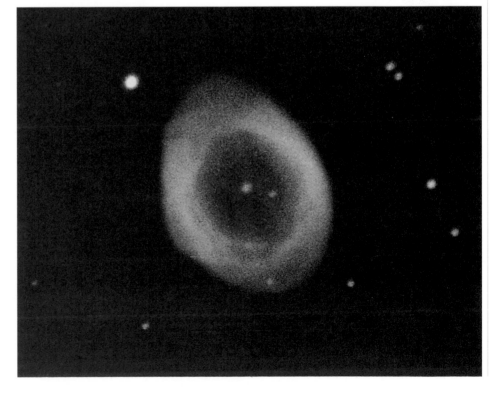

RIGHT: The Ring Nebula, a planetary nebula.

TELESCOPE SIGHTS   EAGLE NEBULA (M16)

DESCRIPTION This emission nebula and embedded star cluster are just visible to the naked eye, and easy to spot in binoculars. A small telescope reveals a handful of stars surrounded by a soft glow coming from the clouds of gas. Use 150x or so and try looking out the corner of your eye—the nebulosity will be easier to see. In medium or large scopes the view is significantly richer and includes many more cluster stars. This whole region is littered with actively forming stars.

VISIBILITY The Eagle can be seen from everywhere except northern high latitudes.

CONSTELLATION Serpens

APPARENT MAGNITUDE +6.0

APPARENT SIZE 20 arcminutes

ACTUAL SIZE 40 light-years

DISTANCE 7,000 light-years

LEFT: Like many regions of dust and hydrogen gas, the Eagle Nebula has a star cluster at its heart.

TELESCOPE SIGHTS   TRIFID NEBULA (M20)

DESCRIPTION This emission/reflection nebula, too faint to be seen by eye, lies in a star-crowded region of the Milky Way. Binoculars show it as a fuzzy patch of light north of the Teapot of Sagittarius. In a small telescope, it looks like a hazy area surrounding two 7th magnitude stars. The nebula's name comes from its appearance in large telescopes and in photos, where three dark dust lanes divide the gaseous southern half—the emission nebula—into three parts. The northern part is a reflection nebula—a cloud of dust reflecting starlight.

VISIBILITY It can be seen from anywhere except the northern high latitudes.

CONSTELLATION Sagittarius

APPARENT MAGNITUDE +6.3

APPARENT SIZE 20 arcminutes

ACTUAL SIZE 47 light-years

DISTANCE 8,000 light-years

LEFT: The Trifid Nebula in star-studded Sagittarius.

TELESCOPE SIGHTS CRAB NEBULA (M1)

DESCRIPTION In a 4 inch (100 mm) telescope, the Crab Nebula is a small oval patch of haze. It hints at an irregular outline, which becomes more evident in 6 inch (150 mm) or larger scopes. The Crab got its name from its appearance in a 36 inch (900 mm) telescope, and the tendrils seen in photos are hard to detect in smaller instruments. The nebula is a supernova remnant—the remains of a star that exploded in AD 1054. The star is now a pulsar spinning 30 times a second, and shining very dimly at 17th magnitude.

VISIBILITY The Crab can be seen from everywhere except antarctic regions.

CONSTELLATION Taurus

APPARENT MAGNITUDE +8.2

APPARENT SIZE 4 x 6 arcminutes

ACTUAL SIZE 10 x 15 light-years

DISTANCE 6,500 light-years

RIGHT: *The Crab Nebula, a supernova remnant.*

TELESCOPE SIGHTS DUMBBELL NEBULA (M27)

DESCRIPTION The Dumbbell's name comes from its wasp-waisted profile, visible in 4 inch (100 mm) or larger scopes. It is a planetary nebula—a cloud of gas shed by a star as it evolves past the red giant stage. The Dumbbell retains its simple hourglass shape in larger scopes, but these pick up a greater area of faintly glowing gas. To see this better, use about 75x magnification and try scanning past the Dumbbell while looking toward the side of the field of view.

VISIBILITY It is visible from locations north of the southern middle latitudes.

CONSTELLATION Vulpecula

APPARENT MAGNITUDE +7.3

APPARENT SIZE 6 arcminutes

ACTUAL SIZE 100 light-years

DISTANCE 1,000 light-years

RIGHT: *Pale green to the eye, the Dumbbell is an easy object for binoculars and telescopes.*

TELESCOPE SIGHTS

WHIRLPOOL GALAXY (M51)

DESCRIPTION In a 6 inch (150 mm) scope at 75x, the Whirlpool and its 9th magnitude companion galaxy NGC 5195 look like two round shining clouds that are just touching. The Whirlpool, the larger of the pair, displays a starlike center. NGC 5195 passed close to the Whirlpool about 400 million years ago, distorting the larger galaxy and starting a burst of star formation within it. In 1994 amateur astronomers discovered a supernova in the Whirlpool Galaxy.
VISIBILITY The galaxies are visible from everywhere except southern high latitudes.
CONSTELLATION Canes Venatici
APPARENT MAGNITUDE +8.3
APPARENT SIZE 8 x 11 arcminutes
ACTUAL SIZE 40,000 light-years
DISTANCE 23 million light-years

LEFT: The Whirlpool Galaxy lies just a few degrees from the star at the end of the Big Dipper's handle.

TELESCOPE SIGHTS

M81

DESCRIPTION This galaxy makes a pair with M82, just ½ degree to the north. In space, M81 and M82 are separated by only about 150,000 light-years. Both are visible even in binoculars. M81 has a bright core and strongly oval shape. It is a spiral galaxy with a large nucleus; M82 is an irregular galaxy. With a 10 inch (250 mm) telescope at 75x, try sweeping across both galaxies to compare their shapes. In 1993 a supernova occurred in M81 and was discovered by an amateur astronomer.
VISIBILITY M81 and M82 can be seen north of the southern middle latitudes.
CONSTELLATION Ursa Major
APPARENT MAGNITUDE +6.8
APPARENT SIZE 20 arcminutes
ACTUAL SIZE 70,000 light-years
DISTANCE 12 million light-years

LEFT: The bright core of the spiral M81 contains most of the galaxy's 250 billion stars.

TELESCOPE SIGHTS   OMEGA CENTAURI

DESCRIPTION You'll think you've found a fuzzy tennis ball when you see Omega Centauri. Easy to spot with the naked eye, this globular cluster looks splendid in scopes as small as 3 inches (75 mm). The cluster responds well both to low powers, which make it stand out against the starry sky, and high powers, which resolve its stars. Its oval shape comes from the movement of its stars, of which there are roughly 2 million. Omega Centauri is the largest, nearest, and most populous of all the Milky Way's known globular clusters.

VISIBILITY The cluster is visible from locations south of the northern tropics.

CONSTELLATION Centaurus

APPARENT MAGNITUDE +3.7

APPARENT SIZE 36 arcminutes

ACTUAL SIZE 180 light-years

DISTANCE 17,000 light-years

RIGHT: The globular star cluster, Omega Centauri.

TELESCOPE SIGHTS   GREAT BARRED SPIRAL (NGC 1365)

DESCRIPTION This barred spiral galaxy can be seen even in 4 inch (100 mm) instruments, but looks much better in larger ones. Use medium power (100x) or more. The bar across its nucleus, from which the arms extend in photographs, is not visible in less than about an 8 inch (200 mm) telescope. In large telescopes, the arms appear as extensions of the elongated nucleus. The Great Barred Spiral appears to be part of a cluster of galaxies in Fornax, but astronomers think it may actually lie in the foreground.

VISIBILITY It can be seen everywhere south of the northern middle latitudes.

CONSTELLATION Fornax

APPARENT MAGNITUDE +9.5

APPARENT SIZE 10 arcminutes

ACTUAL SIZE 100,000 light-years

DISTANCE About 65 million years

RIGHT: NGC 1365, the Great Barred Spiral.

SPACE EXPLORATION

THE TECHNOLOGY OF THE 20TH CENTURY
HAS MADE THE DREAM OF EXPLORING SPACE
A REALITY. ASTRONAUTS HAVE WALKED ON
THE MOON AND LIVED IN SPACE STATIONS,
WHILE ROBOT PROBES HAVE TRAVELED TO
THE OUTER REACHES OF THE SOLAR SYSTEM.
FUTURE EXPLORATIONS MAY WELL
ANSWER THE TANTALIZING QUESTION:
ARE WE ALONE IN THE UNIVERSE?

THE SPACE RACE

Dreams of space travel had spurred thinkers and visionaries for thousands of years, but practical spaceflight began only about 50 years ago in the wake of World War II.

Wartime inventions, notably Germany's *V-2* ballistic missile, showed that rockets were a logical outgrowth of aviation. And the deepening Cold War between the Soviet Union and the United States provided a global arena for the competition known as the space race.

STARTING WITH SPUTNIK

The Soviets launched the first artificial satellite in October 1957. It was called *Sputnik 1*. Passing over the United States several times a day, *Sputnik* gave Americans a profound shock:

it was a great technical achievement, it was larger than any planned American satellite, and it was Russian. A month later *Sputnik 2*, an even larger craft, carried a live dog into space. The first American satellite, *Explorer 1*, was finally launched in January 1958.

The Soviet Union often upstaged the United States in the early years. Working in secrecy, the Soviets could develop projects behind the scenes, conceal failures, and launch spectacular missions timed for maximum publicity. The American program, run by NASA (National Aeronautics and Space Administration), was avowedly civilian and its failures and successes alike occurred in public.

In April 1961 the Soviet Union's *Vostok 1* blasted off for a single orbit of Earth. Aboard was Yuri Gagarin, the first human being to fly in space. Three weeks later the United States launched Alan Shepard on a suborbital flight—his *Mercury* spacecraft traveled beyond Earth's atmosphere but did not go into orbit. The first American to reach orbit, John Glenn, didn't fly until February 1962 and his spacecraft made just three orbits. By that time the Soviets had made a second orbital flight, which completed 17 orbits and lasted more than a day.

The United States appeared to be losing the space race. Few were aware, however, that the *Mercury* craft could be maneuvered by the pilot, while the *Vostok* craft was controlled entirely by autopilot electronics on board and by radio from the ground. The greater sophistication of American spacecraft would become increasingly important as the two programs evolved.

REACH FOR THE MOON

On May 25, 1961, President John Kennedy announced that before 1970 the United States would send a crew to the Moon and return them safely to Earth. To reach this goal, NASA unrolled a systematic plan: the Mercury program was to be followed by Gemini, with a larger craft carrying two astronauts for two weeks, the length of a flight to the Moon. Gemini would also perfect techniques for the rendezvous (meeting) and docking (joining) of spacecraft in orbit. These techniques

LEFT: Valentina Tereshkova was a 26-year-old textile worker and sport parachutist when she became the first woman in space, aboard Vostok 6 *in June 1963.*

BELOW LEFT: Space walks such as Ed White's during the Gemini 4 *mission may have looked like grandstanding stunts, but they taught astronauts important skills for working in space.*

BELOW: Animals always flew first. While the United States favored chimpanzees as test animals, the Soviets used dogs. Because her Sputnik 2 *spacecraft was not designed to return, Laika was put to sleep after a week in orbit.*

ABOVE: Vostok 1 *with Yuri Gagarin aboard was launched on April 12, 1961. Its booster rocket was the same design as the one used for* Sputnik 1. INSET: *Gagarin, a Soviet Air Force lieutenant, made only one spaceflight. He died in a military airplane crash in 1968.*

were important for lunar trips, which would use spacecraft modules that could be linked in orbit and discarded when no longer needed. Following Gemini, the Apollo program would develop a giant booster rocket and three-person spacecraft to take Americans to the Moon.

The Soviets announced no plans, but the Moon was clearly their target also. They developed *Voskhod,* a *Vostok* craft modified to carry a crew of three, and then scored other firsts— the first woman in space, Valentina Tereshkova,

in June 1963, and the first space walk, in March 1965, when Alexei Leonov climbed out of a *Voskhod* spacecraft to float on a tether. Three months later, astronaut Edward White opened the hatch of his *Gemini* craft and did the same.

As 1966 ended, the *Gemini* craft had demonstrated that it could alter its orbit, an ability the *Voskhod* lacked. Moreover, the booster rocket needed for the Apollo program, *Saturn V,* was nearing flight tests, while the Soviet equivalent was scarcely beyond the drawing board.

THE MOON LANDINGS

Early in 1967 the United States and the Soviet Union learned that conquering space had a price beyond dollars and rubles. In January a flash fire swept through an *Apollo* spacecraft, killing astronauts Gus Grissom, Roger Chaffee, and Edward White. Then in April the Soviet Union's Vladimir Komarov was killed when the new *Soyuz* spacecraft crashed on landing. The space race paused, with no manned flights for 18 months.

Missions resumed in October 1968, when the United States tested a redesigned *Apollo* craft in Earth orbit. That same month, the Soviets launched an unmanned *Soyuz,* then a day later sent up a manned one. The two *Soyuz* craft rendezvoused in space but did not dock.

ALMOST THERE

The lunar goal was finally in sight. In December 1968 the gigantic booster rocket *Saturn V* blasted *Apollo 8* into space. After briefly circling Earth, the crew began the long leap to the Moon. Reaching it three days later, they spent a day in lunar orbit before returning to Earth.

The pace quickened. *Apollo 9* tested the lunar lander while orbiting Earth in March 1969. Carrying two of the three crew, this spindly legged craft would separate from the command module and descend to the lunar surface. After spending several hours on the Moon, the two astronauts would lift off in the upper part of the lander and rendezvous with the command module for the journey back to Earth.

In May 1969 the crew of *Apollo 10* flew a full dress rehearsal: they traveled to the Moon, deployed the lunar lander, and took it to within 50,000 feet (15,000 m) of the surface. But the race was not yet over.

ONE GIANT LEAP

On July 3, 1969, the Soviets' lunar booster, luckily unmanned, exploded on launch. Ten days later, preempting the flight of *Apollo 11,* they sent an unmanned probe to the Moon. It was meant to return with samples, but crashed on the Moon's Sea of Crises (Mare Crisium).

Apollo 11, with Neil Armstrong, Edwin Aldrin, and Michael Collins aboard, lifted off on July 16. Reaching lunar orbit on July 20,

Armstrong and Aldrin climbed into their lander and descended to the Sea of Tranquility (Mare Tranquillitatis). Armstrong stepped onto the Moon and uttered "That's one small step for a man, one giant leap for mankind." Aldrin joined him, and the pair spent more than two hours on the surface, placing experiments and collecting rocks. The crew's return to Earth was a huge political and scientific triumph.

LATER LANDINGS

Apollo 11 won the space race, but flights to the Moon continued. *Apollo 12* stayed longer and collected more samples. Then came the drama of *Apollo 13.* A fuel tank exploded en route to the Moon, putting the astronauts in extreme danger. Only heroic efforts and ingenious improvisation saved their lives.

Apollo 14 restored American confidence in lunar flight, and the last three missions— *Apollos 15, 16,* and *17*—targeted areas with spectacular terrain. But critical voices could no longer be ignored. The war in Vietnam and domestic unrest combined to make Moon flights seem extravagant. In December 1972 the Apollo program slipped into history.

While the Moon continues to beckon— and Mars resounds in the imagination—the task of venturing off-planet has been largely turned over to unmanned probes.

ABOVE: Besides opening up the Moon's geological history, lunar samples are still helping scientists better understand Earth and other planets.

RIGHT: Safety worries sent Apollo 11 to the flat Mare Tranquillitatis, mirrored in Edwin Aldrin's visor along with Neil Armstrong. Later Apollo missions explored more interesting sites.
INSET: The Moon's soil has the texture of fine flour mixed with bits of sand and gravel.

BELOW: Against a background of Earth, astronaut David Scott stands in the open hatch of Apollo 9, in 1969.

MEN ON THE MOON

APOLLO 11 • JULY 20, 1969
Neil Armstrong and
Edwin Aldrin
Mare Tranquillitatis
APOLLO 12 • NOVEMBER 19, 1969
Charles Conrad and
Alan Bean
near Mare Cognitum
APOLLO 14 • FEBRUARY 5, 1971
Alan Shepard and
Edgar Mitchell
near Fra Mauro crater
APOLLO 15 • JULY 30, 1971
David Scott and James Irwin
near Montes Apenninus
APOLLO 16 • APRIL 20, 1972
John Young and
Charles Duke
Descartes region
APOLLO 17 • DECEMBER 11, 1972
Eugene Cernan and
Harrison Schmitt
Montes Taurus

PROBES TO THE INNER SOLAR SYSTEM

Manned spaceflight usually gets most of the headlines—and funding. But unmanned probes to the Moon and planets have made a quieter revolution that is still unfolding. Controlled from Earth by radio, and reporting back in a stream of digital data, these robots have changed our view of the solar system's other worlds from enigmatic dots of light in a telescope to places as real as Earth.

LUNAR PROBES

The Moon was the first step outward. The Soviet Union's *Luna 3* probe photographed the Moon's never-before-seen far side in October 1959. The blurry images showed a cratered landscape virtually free of the dark lava flows that mark the near side (see pages 70–73). Many other *Luna* craft and five American *Lunar Orbiters* followed *Luna 3,* all designed to hasten the day of the manned landing (see page 144).

In 1994 the United States returned to the Moon with the *Clementine* probe, which spent two months surveying the Moon's composition and topography from pole to pole. *Clementine* also located possible ice deposits at the south pole and measured South Pole-Aitken Basin, the largest known impact basin at 1,500 miles (2,500 km) wide and 7 miles (12 km) deep.

VENUS AND MERCURY

The first probe to reach another planet was the American *Mariner 2,* which flew past Venus in 1962. While *Mariner* could not see through the clouds shrouding the planet, its instruments detected the hot surface and heavy atmosphere. Venus soon became a favorite target for Soviet

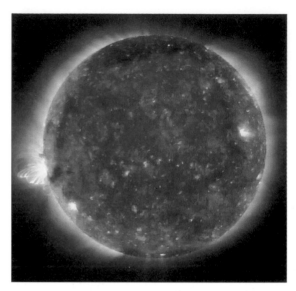

spacecraft, with a series of *Venera* probes taking photos from its surface, and two *Vega* balloon probes analyzing the atmosphere. To create a global view of Venus' surface, scientists turned to radar. A series of missions that included the Soviet *Veneras 15* and *16* (1983) and the United States' *Pioneer Venus* (1978 to 1992) and *Magellan* (1990 to 1994) probes mapped the volcanic landscape in extensive detail (see pages 64–65).

Inward from Venus lies Mercury. Its only visitor has been the American *Mariner 10,* in 1974 and 1975. Because of its trajectory, *Mariner 10* could photograph only about half of Mercury's surface. But that was enough to show the planet's resemblance to the Moon, with impact craters and lava flows (see page 63).

MISSION TO MARS

The planet that has drawn the most interest is Mars, with 20 missions launched toward it. The United States' *Mariner 4* (1965) gave the first close-up look at the Red Planet. It revealed many craters, a blow to those who envisioned Mars as more Earth-like. But *Mariner 9* (1971) found dry riverbeds, again startling scientists. Then came the *Viking* probes (1976), an ambitious—and fruitless—attempt to find life using two automated laboratories that landed where streams might have fed organisms in the soil. Most recently, in 1997, *Mars Pathfinder* and *Global Surveyor* arrived to continue geological surveying. (See also page 79.)

LEFT: Several solar missions have provided astronomers with detailed views of our star. A joint European and American project, the Solar and Heliospheric Observatory (SOHO), *produced this ultraviolet image of the Sun's hot corona, or atmosphere. SOHO has also found a number of Sun-grazing comets.*

ABOVE: This 1968 Soviet postage stamp commemorates Venera 4. *In 1967 the spacecraft had made the first sampling of Venus' atmosphere, discovering that it is much denser than Earth's and almost entirely made up of carbon dioxide.*

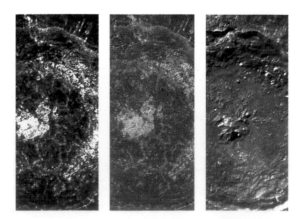

LEFT: As part of its global survey of the Moon, the Clementine *probe produced data for these images of the crater Tycho. The center image is colored to identify different types of rock. The image at left emphasizes iron-rich material, while the one at right enhances the natural colors in the crater.*

ON THE WAY OUT

Beyond Mars lies the asteroid belt, a target of opportunity for spacecraft passing through to the outer planets. The United States' *Galileo* craft was bound for Jupiter when it flew past asteroids 951 Gaspra (in 1991) and 243 Ida (1993). The latter had a 1 mile (1.6 km) wide moon, since named Dactyl. In 1997 the American *Near-Earth Asteroid Rendezvous* spacecraft traveled past asteroid 253 Mathilde, continuing the geological census of these varied minor worlds. (See also pages 80–81.)

ABOVE: Magellan *was launched from the space shuttle* Atlantis *in 1989. Arriving at Venus in 1990, the probe used radar to peel away the planet's clouds.*

RIGHT: Sojourner, *the rover carried by Mars Pathfinder, was designed to sample a former water channel that may contain rocks from other parts of Mars.*

PROBES TO THE OUTER SOLAR SYSTEM

Sending a probe to Mars or Venus is relatively simple, but the outer solar system poses a tougher target, with flights taking years rather than months. Because of the great distances, the spacecraft must operate automatically—radio commands from Earth can take an hour to reach even Jupiter, the closest gas giant.

TWO PIONEERS

The few missions that have ventured beyond the asteroid belt have been American. The first was *Pioneer 10,* which arrived at Jupiter in 1973. It drew a detailed portrait of the giant world, mapping its powerful magnetic field and creating the first close-up images of its roiling atmosphere. *Pioneer 11* followed a year later, confirming and extending the findings.

On leaving Jupiter, *Pioneer 11* took aim at Saturn, flying past it in 1979. The spacecraft mapped the planet's cloud belts, sampled its magnetic environment, and found new rings.

Both *Pioneer* craft were traveling quickly enough to escape the solar system. Outward bound forever, each craft carries a plaque with a greeting from Earthlings to whomever might find it in the depths of interstellar space.

BELOW: *When the two Voyager craft took a close look at Saturn's rings, scientists got a jolt. They discovered that their tidy Earth-based picture of three rings dissolved into a tissue of ringlets within ringlets, all moving in complex mathematical patterns.*

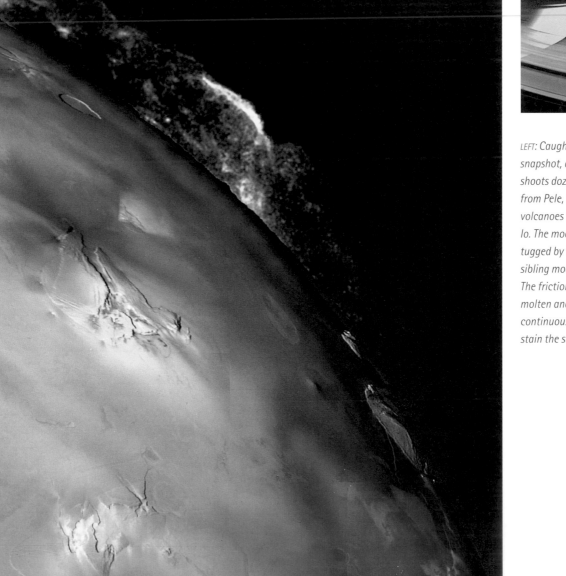

LEFT: *Caught in a Voyager snapshot, a plume of sulfur shoots dozens of miles high from Pele, one of the many volcanoes on Jupiter's moon Io. The moon is constantly tugged by the gravity of its sibling moons and Jupiter. The friction keeps Io's interior molten and causes almost continuous eruptions, which stain the surface with sulfur.*

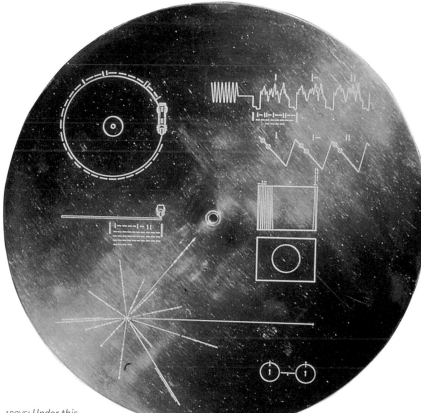

ABOVE: Under this gold-plated cover (showing Earth in relation to nearby pulsars) is a recording with greetings from world leaders, samples of our music, and photographs of terrestrial scenes. A copy of the record was placed on both Voyagers as a message to any civilization that may find the craft.

A PAIR OF VOYAGERS

Pioneer's reconnaissance was soon eclipsed by Voyager, a project involving two sophisticated spacecraft. Both craft reached Jupiter in 1979 and then moved on to Saturn, *Voyager 1* arriving in 1980 and *Voyager 2* in 1981. Firing photos back to Earth every 48 seconds, they revolutionized the study of the two planets.

Among Voyager's many findings at Jupiter were details of complex storms in its atmosphere and volcanic eruptions on its moon Io. At Saturn the craft discovered that the rings consist of innumerable ringlets and that dark organic compounds cover one side of the moon Iapetus. *Voyager 2* traveled on to Uranus (1986) and Neptune (1989). It was the first spacecraft to survey these worlds, which appear as no more than tiny disks in Earth-bound telescopes.

Besides sending back remarkable views of the four planets, the Voyager project roughly doubled the number of known moons at each planet. Saturn's rings had been known about for hundreds of years, but it took Voyager to reveal the ring systems of Jupiter, Uranus, and Neptune. (See also pages 84–95.)

RIGHT: Galileo has investigated Jupiter's Great Red Spot (seen here in false color), a long-lived storm more than twice the size of Earth. The craft found giant thunderstorms (insets) feeding the 300 mile per hour (500 km/h) winds that swirl around the Spot.

GALILEO AND JUPITER

To follow up on Voyager's study of Jupiter, the United States developed *Galileo*. The *Pioneers* and *Voyagers* were flyby spacecraft, grabbing snapshots of each planet as they whizzed past. But *Galileo* was designed to spend several years orbiting Jupiter, like a new kind of inquisitive moon moving among the natural ones.

Galileo reached Jupiter in 1995. On arrival it shot a probe straight into Jupiter's atmosphere. As the probe fell through the clouds, its instruments sampled gases and measured wind speeds and temperatures. The main craft then settled into orbit around Jupiter and began its survey. *Galileo* revisited some of the perplexing questions raised by Voyager, including: What drives Jupiter's clouds? and, Does a global ocean lie beneath the icy skin of the moon Europa? (See also pages 84–87.)

COMET PROBES

Although comets dwell in the outer reaches of the solar system (see page 98), sending probes out there is impractical. Thus when comet Halley flew past Earth in 1986, it was greeted by a small fleet of spacecraft, including Europe's *Giotto,* the Soviet Union's *Vega 1* and *2*, and Japan's *Sakigake* and *Suisei*. *Giotto* flew closest and, before dust particles blinded its cameras, snapped photos of Halley's dark nucleus.

SPACE SHUTTLES AND STATIONS

As *Apollo 11* landed on the Moon, the United States was designing the first reusable spaceship. Dubbed the space shuttle, this craft could orbit Earth for two weeks at a time while carrying up to seven astronauts and a cargo of more than 30 tons (27 tonnes).

FLY IT AGAIN

The first space shuttle, *Columbia,* was launched in April 1981. Three more shuttles soon joined the fleet, and as launch followed launch, spaceflight became routine. Or so it seemed until the disaster of January 1986, when the shuttle *Challenger* exploded 73 seconds after liftoff, tragically killing its crew.

After being grounded for two years, shuttles returned to flight in September 1988, launching the *Magellan* and *Galileo* probes to Venus and Jupiter and putting the Hubble Space Telescope into Earth orbit (see page 26).

The shuttle is an all-purpose "space truck." It delivers satellites into orbit and retrieves them for repair, and can serve as the first stage for probes to other planets. The shuttle has also carried *Spacelab,* a manned space laboratory that enables scientists to conduct experiments in a weightless environment. But the most important role for the shuttle is building and supplying the *International Space Station.*

SALYUT TO MIR

Originally, manned space stations in orbit around Earth were seen as the first step in any Moon flight, but the Cold War scrambled priorities. In the end, the Soviet Union was the first to develop a space station, launching seven stations, all named *Salyut,* between April 1971 and April 1982. Crew were transported to and from the station by a *Soyuz* craft. Sadly, the first crew of *Salyut 1* died when their craft lost air pressure as they returned to Earth.

Over the course of the Salyut program, the Soviets pioneered the study of long-term weightlessness. The flight plan was for a crew to spend up to a year in the station, while other crews (including some non-Russians) arrived for short visits. Besides conducting medical and psychological tests, the crews made astronomical observations and studied Earth.

Following the Salyut series came a more ambitious craft named *Mir,* which is still in orbit. It has been enlarged several times, with each addition focusing on special tasks, such as studying Earth's environment or observing stars and galaxies. But while *Mir* is still in operation, it is an old spacecraft and its obsolete systems are wearing out.

SKYLAB

When Moon landings were cut short, the United States modified a leftover *Saturn V* booster, turning its upper stage into roomy living quarters and laboratories and renaming it *Skylab.* The craft was launched into Earth orbit, where it was inhabited by three crews between May 1973 and February 1974.

As a "house-in-space," *Skylab* might have formed the nucleus of an American space station a quarter-century ago. But its orbit decayed faster than anticipated and there was no way to reboost it. On July 11, 1979, it reentered Earth's atmosphere and was destroyed.

In recent years the American concentration on the shuttle and the Russian focus on *Mir* have converged in preparation for the new *International Space Station,* being built with hardware and ideas from many countries and due for its first inhabitants in 1999. Russian cosmonauts have flown on the shuttle and American astronauts have spent months aboard *Mir,* all of them gaining valuable experience.

LEFT: *When astronauts begin assembling the* International Space Station, *they will work with devices such as the shuttle's remote-manipulator arm, used here by astronaut Bruce McCandless.*

ABOVE: *Astronaut Robert Gibson greets cosmonaut Vladimir Dezhurov during the first docking of a space shuttle at* Mir, *in July 1995.*

RIGHT: *The shuttle* Discovery *was launched in 1983. The winged orbiter rides on a red fuel tank, which is flanked by two white solid-fuel boosters. The fuel tank and boosters are jettisoned when they burn out.*

ABOVE: Mir *floats above the Tasman Sea, photographed from the shuttle* Atlantis *just before the third shuttle-Mir docking, in March 1996.*

THE FUTURE IN SPACE

As Cold War tensions faded, a new sense of cooperation made spaceflight international, and governments began to collaborate.

The next big project is the *International Space Station.* Largely an initiative of the United States, the station will use major components from 14 other countries, including Russia, Japan, France, Germany, the United Kingdom, and Canada. The first part will be put into orbit in 1997, with other modules added later. An international crew will arrive in January 1999.

Measuring 356 by 290 feet (110 by 90 m), the station will provide a home for up to seven crew members, who will conduct a range of scientific experiments. The station will also demonstrate the effects of extended space travel on humans, which should help in the planning of manned expeditions to the Moon and Mars, the two most feasible targets.

REMOTE CONTROL

In the realm of unmanned probes, the tighter budgets of recent years have discouraged large, expensive missions. The new theme is to fly smaller spacecraft far more often.

Among the upcoming American missions are new infrared and ultraviolet observatories

and the Advanced X-Ray Astrophysics Facility, a satellite observatory to explore high-energy processes in the universe. The Hubble Space Telescope will continue to receive regular upgrades. (See also pages 26–27.)

A mission called Genesis will collect samples from the solar wind and return them to Earth for analysis, giving scientists data to test theories of how the Sun and planets formed.

Lunar Prospector, launched in early 1998, will map the chemical composition of the Moon's surface and its magnetic and gravity fields. It should also determine whether the shadowed craters near the poles contain any water-ice.

The *Mars Global Surveyor* left Earth in 1997. Its orbiter and lander will study Mars' weather, analyze its water and carbon dioxide, and search for evidence of climate change. The European *InterMarsNet* is an orbiter with a collection of landers designed to study the internal structure of Mars. Arriving in 2004, the landers would operate for one Martian year.

ASTEROIDS, COMETS, AND PLUTO

The *Near-Earth Asteroid Rendezvous* mission visited minor planet 253 Mathilde in 1997, and is now heading for 433 Eros. In 1999 the craft will go into orbit around the asteroid.

For comets, the United States will fly a Comet Nucleus Tour to photograph three comets at close range in the first decade of the new century. Also, its *Stardust* probe to

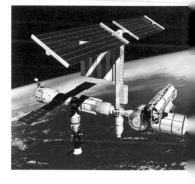

TOP: *This illustration shows what a manned outpost on the Moon might look like. The long inflatable structure represents a permanent lunar habitat for 12 people.*

ABOVE: *The modular approach to building the* International Space Station *has been dictated by what launch vehicles can carry, but it will make upgrading the station easy.*

LEFT: *To perform for months in the Space Station, crews will need living quarters designed for weightlessness, including showers that are equipped with handholds and footholds.*

ABOVE: *In this illustration, the* Huygens *probe has landed on Titan, Saturn's largest moon. Due to arrive in 2004,* Huygens *will sample Titan's atmosphere on its descent. If the probe survives the landing, it will also study surface conditions.*

Comet Wild 2 will bring back particles from the comet's tail in 2006. The Europeans plan a mission called *Rosetta* to rendezvous with Comet Wirtanen in 2011. *Rosetta* will travel with the comet for a year and deposit a lander craft on its surface.

In the outer planets, the *Cassini* mission to Saturn is broadly similar to *Galileo*'s at Jupiter (see page 149). Arriving in 2004, *Cassini* will land a probe on the smog-shrouded moon

Titan. The orbiter will then spend several years studying Saturn's clouds and storms, its multitudinous rings, and its ice-and-rock moons.

One planet still awaits its first visit—Pluto. The United States has designed a low-cost mission to be launched in 2001 that would arrive at Pluto in 2013. After flying past Pluto, the probe would continue into the Kuiper Belt (see page 98), and then out of the solar system toward interstellar space.

ARE WE ALONE?

No one knows if humanity is alone in the universe or not. (And, as the joke goes, either way it's a mighty sobering thought.) But when scientists examine the evidence of how life began on Earth, the process does not look in any way special or unusual.

THE ORIGINS OF LIFE

Clearly, life began a long, long time ago. The oldest rocks on Earth, which date from more than 3.9 billion years ago, are of a kind that could never preserve fossil traces of life. But sedimentary rocks only slightly younger contain amounts of carbon that may have come from living organisms, and fossil bacteria have been identified in rocks that are 3.5 billion years old.

In any case, life arose during a period in Earth's history completely unlike today. Comets and asteroids were often colliding with our planet. They brought much devastation—witness the Moon, whose surface dates from this time. Area for area, Earth probably bore as many impact scars as the Moon does now.

But impacts, especially from comets, also brought water and compounds rich in carbon.

Carbon was especially important. It forms the chemical basis of life today, and it is the most versatile of the elements, building many compounds that life depends upon. Combined with the byproducts of Earth's volcanic activity these ingredients created the first oceans, a rich stew out of which life arose, aided by the energy carried in sunlight.

Scientists believe that life may have started several times, being destroyed again and again by violent impacts. At some point, however, the rate of impacts slackened enough to let life endure, and from that day forward it evolved.

EXTRATERRESTRIAL LIFE

Is Earth alone in having life? Impact craters cover the surfaces of every planet and moon in the solar system, so it's conceivable that comets triggered some form of life elsewhere.

In 1996 scientists discovered possible fossils in a meteorite from Mars. For its first billion years, Mars' geological evolution paralleled Earth's (see page 76). Life could have arisen when Mars was warmer and wetter than it is today—and then died out as conditions became too hostile.

BELOW: A meteorite from Mars, known as ALH84001, drew worldwide attention in 1996 when scientists announced it contained evidence of early Martian life. The evidence included micro-structures (colored blue in this image) that resemble bacteria.
INSET: Earth's bacteria, such as Haemophilus influenzae, are about 100 times larger than the Martian micro-structures.

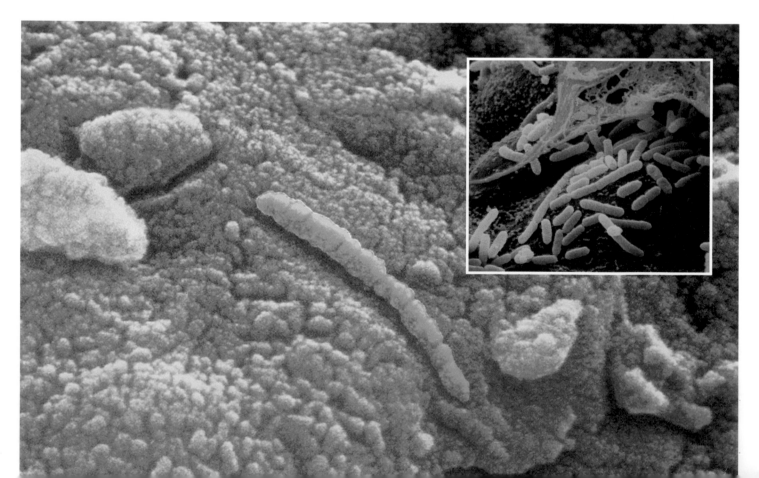

ABOVE: The icy crust of Jupiter's moon Europa looks a bit like arctic ice floes on Earth. What lies under the icy shell? Cracks (colored red in this image) show where the crust has broken. Water, perhaps laden with dust, has filled the gaps and then frozen. Scientists suspect that the crust covers a global layer of water or soft ice where life might exist.

A much more remote possibility is that there might be life in the global layer of water or slush that scientists believe underlies the ice floes covering Jupiter's moon Europa. The energy to drive life could come from weak sunlight filtering through the ice or from heat generated by the tides created on Europa by its sibling moons and Jupiter. This might create an environment where life could exist as it does in the depths of Earth's oceans.

Europa is certainly no garden spot, but it's worth recalling that life on Earth is extraordinarily abundant and tenacious. Any ecological niche that affords even the slightest foothold has its inhabitant. It would therefore be only prudent to assume that all life has this

characteristic, and thus we shouldn't write off even the unlikely locations too quickly.

And why stop with the Sun's family? The universe has billions of galaxies, each containing perhaps 100 billion suns, and some of these suns are certainly orbited by planets (see page 44). Astronomers know that chemical elements behave in the most distant galaxies just as they do here on Earth. Two atoms of hydrogen combined with one of oxygen yields water—everywhere. And carbon would be just as chemically versatile and suitable for life in the Andromeda Galaxy as it is here.

While the search continues for other life, it's worth letting the possibilities trickle through your mind as you look up into a star-strewn sky.

GLOSSARY

ACTIVE GALAXY A galaxy with a central black hole that is emitting lots of radiation.

APERTURE The diameter of a telescope's main light-collecting optics, whether a lens or a mirror. Also, the diameter of a binocular lens.

ARCMINUTE A unit of angular measure equal to 1/60 of a degree; the Moon and Sun are about 30 arcminutes across.

ARCSECOND A unit of angular measure equal to 1/60 of an arc-minute; Jupiter averages some 44 arcseconds across.

ASTRONOMICAL UNIT (AU) The average distance between Earth and the Sun, about 93 million miles (150 million km).

ATMOSPHERE The layer of gases attached to a planet or moon by gravity.

BIG BANG The explosion of a small, very hot lump of matter about 15 billion years ago that marked the birth of the universe, according to the current theory of the universe's origin.

BLACK HOLE A massive, compact object so dense that no light or other radiation can escape from inside it.

CONSTELLATION One of the 88 official patterns of stars that divide the sky into sections.

DEGREE A unit of angular measure equal to 1/360 of a circle; your thumb held at arm's length spans about 2 degrees.

ELECTROMAGNETIC SPECTRUM The full range of radiation produced by nature; runs from high-energy gamma rays to very long radio waves.

FALSE-COLOR An enhancement technique used in image processing to make subtle differences stand out.

FOCAL LENGTH The distance between the main lens or mirror of a telescope and the point where the light from it comes to a focus.

GAMMA RAYS Radiation shorter than X rays.

GAS-GIANT PLANET A planet whose composition is dominated by hydrogen: Jupiter, Saturn, Uranus, and Neptune.

IMPACT CRATER The round, bowl-like scar left on the surface of a moon or planet when it is struck by a meteorite.

INFRARED (IR) Radiation with wavelengths just longer than visible light.

LIGHT-YEAR The distance light travels in a year, about 6 million million miles (9.5 million million km).

M OBJECTS Star clusters, nebulae, and galaxies in the Messier list, compiled by 18th-century comet-hunter Charles Messier to create a roster of objects that resembled comets.

MAGNETIC FIELD The region of space in which a celestial object exerts a magnetic force.

MAGNITUDE The unit of brightness for celestial objects. Apparent magnitude describes how bright a star looks from Earth, while absolute magnitude is its brightness if placed at a distance of 32.6 light-years.

MICROWAVE Radiation with wavelengths measured in millimeters.

NEBULA A cloud of gas or dust in space; may be either dark or luminous.

NGC OBJECTS Galaxies, star clusters, and nebulae listed in the *New General Catalogue* of J.L.E. Dreyer, published in the late 19th century.

NUCLEUS The central core of a galaxy or comet.

OPTICAL Radiation with wavelengths detectable by the human eye.

ORBIT The path of an object as it moves through space under the control of another's gravity.

PULSAR An old, rapidly spinning star that flashes periodic bursts of radio (and occasionally optical) energy.

QUASAR A quasi-stellar object or radio source whose spectrum shows a high velocity of recession from Earth; quasars are thought to be the active nuclei of very distant galaxies.

RADAR ASTRONOMY The study of solar-system objects using a radio telescope to bounce signals off their surfaces.

RADIATION The means by which energy travels through space; it has characteristics of both waves and particles.

RADIO Radiation measured in centimeters and longer.

RED GIANT A large, cool, red star in a late stage of its life.

SATELLITE Either a moon or a spacecraft in orbit around a planet.

SUPERNOVA The explosion of a star in which it blows off its outer atmosphere and briefly equals a galaxy in brightness.

TECTONIC Geological movements on a planet or moon, driven by forces that produce folding and faulting.

TERRESTRIAL PLANET A planet whose composition is mainly rocky: Mercury, Venus, Earth, and Mars.

ULTRAVIOLET (UV) Radiation with wavelengths just shorter than visible light.

VOLCANIC Geological activity driven by the internal heat of a planet or moon.

WAVELENGTH The distance between two successive waves of energy passing through space.

WHITE DWARF The small, very hot remnant of a star that has evolved past the red giant stage.

X RAYS Radiation with wavelengths between ultraviolet and gamma rays.

INDEX

Bold page numbers indicate main references while figures in italics indicate photographs, illustrations, and captions.

CONTRIBUTORS AND PICTURE CREDITS

CONTRIBUTORS

ROBERT BURNHAM A science writer specializing in astronomy and earth science, Robert Burnham is a former editor-in-chief of *Astronomy* magazine and the author of *Comet Hale-Bopp: Find and Enjoy the Great Comet* and *The Star Book*. He has been an amateur astronomer since the 1950s, mainly observing the Moon and planets with his backyard telescope. He also enjoys following developments in cosmology.

GABRIELLE WALKER As Physical Science Editor of *New Scientist* magazine, Gabrielle Walker is responsible for the magazine's features in all of the physical sciences, but specializes in astronomy and earth science. A former editor at *Nature* magazine, she first worked as a freelance science writer while completing a Ph.D. in Chemical Physics at Cambridge University.

PHOTOGRAPHIC CREDITS

A = Auscape International; AAO = David Malin/Anglo-Australian Observatory; AF = Akira Fujii; ASP = Astronomical Society of the Pacific; AURA = Association of Universities for Research in Astronomy; Bridgeman = The Bridgeman Art Library, London; Granger = The Granger Collection, New York; IS = Image Select; JPL = Jet Propulsion Laboratory; MIC = Meade Instruments Corporation; Mendillo = Mendillo Collection of Antiquarian Astronomical Prints; NC = Newell Color; NOAO = National Optical Astronomy Observatories; NRAO = US National Radio Astronomy Observatory; PE = Planet Earth Pictures; TPL = The Photo Library, Sydney; SF = Space Frontiers Collection, London; SM = Science Museum/Science and Society Picture Library; SPL = Science Photo Library; STScI = Space Telescope Science Institute; TS = Tom Stack & Associates; UCO = University of California Observatory; USGS = US Geological Survey; WO = Weldon Owen

t = top, b = bottom, l = left, r = right, c = center, i = inset

front jacket Luke Dodd/SPL/TPL back jacket NASA; i AF 1 Luke Dodd/SPL/TPL 2 NC/AURA/STScI/NASA 3 Michael Terenzoni 4t SM; c RGO/SPL/TPL; b Dr J Durst/SPL/TPL 5t Frank Zullo/Photo Researchers/TPL; b NASA/SPL/TPL 6–7 European Space Agency/SPL/TPL 7i SM 8 I.M. House/Tony Stone Images/TPL 9t NASA; b SM 10t Granger; b Tony Stone Worldwide/TPL 11 Victoria and Albert Museum, London/Bridgeman 12t Museo del Banco Central de Ecuador/D.Donne Bryant; b British Museum/Robert Harding Picture Library 13t Granger; b William Macquitty/Camera Press/Austral International 14 Musee Conde, Chantilly/Giraudon/Bridgeman 15t Granger; bl Granger; bc Bulloz; br Louvre, Paris/Bridgeman 16 Granger 17t SM; c SM; b Granger 18bl Hulton-Deutsch/TPL; br IS 19t Ann Ronan/IS; b SPL/TPL 20 Roger Ressmeyer/Corbis; i Steve Northup/Black Star/Colorific/PE 21t Granger; b AAO 23t NASA/JPL/SPL/TPL; bl Leiden Observatory/SPL/TPL; br Max-Planck-Institut/SPL/TPL 24t Tim Acker/A; c Dr Stephen Unwin/SPL/TPL; b Ann Ronan/IS

25t SF/PE; b Tony Craddock/SPL 26l SF/PE; r NASA/SPL/TPL 27l AURA/STScI/NASA; r Max-Planck-Institut/SPL/TPL 28–29 NRAO/SPL/TPL 29i RGO/SPL/TPL 30 NC/AURA/STScI/NASA 31 NC/AURA/STScI/NASA 33 NASA/SPL/TPL 34l AAO; r NC/AURA/STScI/NASA 35 NC/AURA/STScI/NASA 36 AURA/STScI/NASA 37t John Chumack/TPL; bl SPL/TPL; br AAO 38t AAO; b Bill and Sally Fletcher/TS 39 NC/AURA/STScI/NASA 40t STScI/NASA/SPL; b AAO 41 NASA/SPL/TPL 42 NASA/SPL/TPL 43t Max-Planck-Institut/SPL/TPL; bl Dr Rudolph Schild/SPL/TPL; br NASA/SPL/TPL 44 SPL/TPL 45t Lynette Cook/SPL/TPL; b NASA/SPL/TPL 46t USNO/TSADO/TS; bl SPL/TPL; br NASA/SPL/TPL 47t AAO; b NOAO 48 Bill and Sally Fletcher/TS; 48–49 NASA/SPL/TPL 49 Luke Dodd/SPL/TPL 50tl NC/STScI/NASA; tr Dr Rudolph Schild/SPL/TPL 51t SPL/TPL; c STScI/NASA; STScI/NASA/SPI/TPL 52 PE 53 AAO/Royal Observatory Edinburgh; i Dr K Milne/D.Parker/SPL/TPL 54–55 NASA 55i Dr J Durst/SPL/TPL 56 NASA 58t NASA/TSADO/TS; bl NASA/SPL/TPL; br Geoff Tompkinson/SPL/TPL 59 JISAS/LOCKHEED/SPL/TPL 60–61 NASA 61 NASA/TS 62t Fred Espenak/SPL/TPL 63tl Dr Michael J Ledlow/SPL/TPL; r NASA/Astrovisuals; bl JPL/TSADO/TS 64tr Fred Espenak/SPL/TPL; bl NC/NASA 65tl tr NASA/Astrovisuals; bl SPL/NASA/TPL; br David P Anderson/NASA/SPL/TPL 66t Ferrero-Labat/A; b John McKinnon/Auscape 67b Soames Summerhays/TPL 68 Steve McCurry/Magnum 69 Kas Mori/The Image Bank; i NASA 70 National Geographic/Horizon 71 Alan Dyer 72 NASA 73 UCO/Lick Observatory 74t AF; c Fred Espenak/SPL/TPL 75 AF; i David Miller 76t NASA; b Jack Chertok Television Inc./Len Peltier 77t Kevin Kelley/TPL; b USGS/TSADO/TS 78 USGS/NASA/TSADO/TS 79tl North Wind Picture Archives; tr NASA/SPL/TPL; bl JPL/TSADO/TS 80tl NASA; tr NASA/SPL/TPL; br Jayaraman & Dermott/SPL/TPL 81t The John Hopkins University Applied Physics Laboratory; b B.Freeman/Australian Picture Library 82 Royal Observatory, Edinburgh/TPL 83t John Sanford/SPL/TPL; cl Kate Lowe/Nature Focus/Australian Museum; b Jean-Paul Ferrero/A 84 NASA/SPL/TPL 85 NASA; i NASA/SPL/TPL 86 NASA/SPL/TPL 87t Astrovisuals/NASA; b NC/NASA 88 NASA 89tl Royal Observatory Edinburgh/SPL/TPL; tr NASA; b NASA 90t NASA; b JPL/TSADO/TS 91t NASA; bl Peter H.Smith/University of Arizona Lunar and Planetary Laboratory and NASA; br Photo Research International/TPL 92tl NASA/JPL/TS; tr SF/PE 93t NASA; b NASA; i SPL/TPL 94tr NASA/SPL/TPL; cl SPL/TPL; cr Mary Lea Shane Archives/Lick Observatory/ASP slide set "Astronomers of the Past" 95 NASA; i NASA/SPL/TPL 96t NC/NASA; b Corbis Bettmann/Australian Picture Library 97tl NC/NASA; tr Alan Stern/Southwest Research Institute, Marc Buie/Lowell Observatory, NASA and ESA/ASP slide set "HST: What A View" 98l Dennis di Cicco/Peter Arnold, Inc./A; r Jane Luu/U.C. Berkeley and David Jewitt/U. Hawaii/NASA 99t John Chumack/TPL; b AF 100–101 NASA

101i Frank Zullo/Photo Researchers/TPL 102 John Chumack/Galactic Images/TPL 103t Jerry Lodriguss; b Michael Terenzoni 104 Stapleton Collection/Bridgeman 105t AF; b WO 106 AF 108 Granger 109t Mendillo; b SM 110t Mendillo; b Lambeth Palace Library/Bridgeman 111t British Library, London/Bridgeman; b Granger 112t Granger; b British Library, London 113t Granger; b Mendillo 114t Granger; b SM 115t Mendillo; b Granger 116 Jerry Schad 117 AF 118t Alan Dyer; b AF 119 David Miller 120t Jack Finch/SPL/TPL; b Jerry Schad 121t David Miller; b AF 122t AF; bl Oliver Strewe/WO; br AF 123t Jerry Lodriguss; b WO 124t Michael Terenzoni; b AF 125t AF; b Luke Dodd 126t Lee C.Coombs; b SPL/TPL 127t AF; b Bill and Sally Fletcher 128t Bill and Sally Fletcher; b Luke Dodd 129t NOAO/TS; b AF 130 MIC 131t Oliver Strewe/WO; bl Orion Telescopes and Binoculars; br MIC 132t WO; b AF 133t Jerry Lodriguss; b WO 134 AF 135t National Solar Observatory/ASP slide set "Secrets of the Sun;" b Kim Gordon/SPL/TPL 136t AAO; b Jerry Lodriguss 137t NOAO/TSADO/TS; b IAC Photo from plates taken with the Isaac Newton telescope by David Malin 138 Bill and Sally Fletcher 139t Luke Dodd/SPL/TPL; b AAO 140–141 Mark M. Lawrence/TSM/Stock Photos Pty Ltd 141i NASA/SPL/TPL 142tl Novosti, London; bl NASA; br Novosti, London 143 Novosti, London 144 NASA/SPL/TPL 145 NASA 146tl European Space Agency/SPL/TPL; tr Novosti, London; b SF/PE 147t SF/PE; b NASA 148l NC/NASA; r NASA/SPL/TPL 149t NASA/JPL/TSADO/TS; b SF/PE 150tl International Photographic Library; tr SF/PE; br NASA/SPL/TPL 151 NASA 152tr NASA; cr NASA; bl NASA/SPL/TPL 153 European Space Agency/SPL/TPL 154b NASA/SPL/TPL; bi Dr Tony Brain/SPL/TPL 155 NASA/SPL/TPL

ILLUSTRATION CREDITS

Nick Farmer/Brihton Illustration: 34, 36, 59, 67, 71, 74, 84.
Chris Forsey: 35, 57, 60, 61, 80, 97, 98, 104, 122, 131.
Robert Hynes: 22, 32, 41, 49, 68.
Avril Makula: 44.
Wil Tirion: star charts 106–115.

CAPTIONS

1 The constellation of Crux, the Southern Cross.
2 Detail of the Lagoon Nebula, photographed by the Hubble Space Telescope.
3 Comet Hale-Bopp in sky above Tucson, Arizona.
6–7 The *SOHO* satellite gathered the data for this graph, which reveals details of the Sun's internal composition and structure. Inset: The constellation of Aries, the Ram, on a 1533 celestial globe.
28–29 Composite optical and radio image of radio galaxies NGC 7018 and NGC 7016. Inset: Optical image of Supernova 1993J in spiral galaxy M81.
54–55 *Magellan* radar image of Ushas Mons, a volcano on Venus. Inset: Solar corona during a total eclipse.
100–101 The Great Nebula in Orion (M42). Inset: Star trails in Arizona.
140–141 Launch of space shuttle *Discovery*. Inset: Astronaut Bruce McCandless during a space walk.